PEST CONTROL IN

TROPICAL ROOT CROPS

PANS MANUAL No. 4

CENTRE FOR OVERSEAS PEST RESEARCH
OVERSEAS DEVELOPMENT ADMINISTRATION
LONDON

First edition	1978
Reprinted	1982
Reprinted	1986

ISBN 0 85135 089 5

Printed by Hobbs the Printers of Southampton
for Centre for Overseas Pest Research, College House, Wrights Lane, London W8 5SJ, UK

CONTENTS

YAMS

TARO

FOREWORD

Root crops are important foods for many people in the tropics and subtropics and it was therefore considered appropriate by the Natural Resources Advisers Co-ordinating Committee of the UK Ministry of Overseas Development that the fourth *PANS Manual* should deal with some of these, namely; sweet potato, cassava, yams and taro. These crops are important to subsistence farmers and ODM policy is to provide more assistance in this area of rural development. This *PANS Manual,* like others in the series, is regarded by ODM as an aid operation to developing countries and hence is available free of charge to agriculturalists and agricultural scientists in most developing countries.

This manual contains general chapters on weed control, rodent control and storage followed by sections on control of diseases, nematodes and insect pests of each crop.

Throughout the manual the pesticides involved are referred to by accepted common names and at the end of the manual an appendix is provided in which the common names are listed with some of their trade names. When using pesticides farmers should follow the safety precautions indicated on the label of the product. Concentrated pesticides are dangerous substances and must be treated with care. When the application of pesticides has been completed for the day the farmer should wash himself, his clothes and machinery. Pesticide containers must not be used for any other purpose and should be destroyed. The minimum period between pesticide application and harvest should be stated on the label, if it is not then local advice from the manufacturer's agent or agricultural adviser should be sought. In this manual the minimum interval before harvest indicated in the country from which the recommendation originates has been included, if available; however, the regulations differ and should be checked locally. It must also be remembered that the application of a pesticide may kill beneficial predators and parasites, for example mite predators and if the pest population recovers faster than the predator or parasite populations it may be necessary to treat the crop a second time with the pesticide.

The CIE and CMI Distribution Map numbers given in the insect pest and disease sections refer to the *Distribution Maps of Insect Pests* and the *Distribution Maps of Plant Diseases* published by the Commonwealth Institue of Entomology and the Commonwealth Mycological Institute, respectively. The scientific names of the majority of insects and diseases mentioned in the text have been checked with these Institutes, and are given in full, with authorities and synonyms in check lists in the Appendix.

The sections on weeds, rodents, storage, nematodes, diseases and insects were written by L. Kasasian, E. J. Taylor, R. H. Booth, J. Bridge, S. C. A. Cook and A. Ward respectively, technical editing was by S. L. Mercer and the artwork was by H. R. Broad and L. Huddleston.

During the preparation of *Pest Control in Tropical Root Crops* PANS staff have been greatly helped by Mr D. G. Coursey, Plant Food Commodities Department, Tropical Products Institute, London, UK and Dr D. E. Shaw, Department of Primary Industry, Konedobu, Papua New Guinea and further assisted by advice and publications from many organisations and individuals and would like to acknowledge the help of the following:

Mr W. G. L. Austin, Plant Protection Division, ICI Fernhurst, Haslemere, UK.
Dr E. W. Baker, Systematic Entomology Laboratory, Beltsville, USA.
Dr A. Bellotti, Centro Internacional de Agricultura Tropical, Cali, Colombia.
Mr S. Bharath, Central Experiment Station, Centeno, Asima, Trinidad.
Miss M. Bolton, Horticultural Research Institute, Pretoria, South Africa.
Dr J. Bradbury, Commonwealth Mycological Institute, Kew, UK.
Dr H. R. Brandes, Plant Science Research, Elanco, Windlesham, UK.
Mr L. L. Cecil, Diamond Shamrock, Cleveland, USA.
Mr A. Charles, Department of Primary Industry, Konedobu, Papua New Guinea.
Professor H. Dempsey, University of Georgia, Georgia, USA.
Dr Dieber, Horticultural Research Institute, Pretoria, South Africa.
Dr J. D. Doll, University of Wisconsin, Madison, USA.
Miss J. M. Donaldson, Horticultural Research Institute, Pretoria, South Africa.
Mr D. C. Drummond, Ministry of Agriculture, Fisheries and Food, Surbiton, UK.
Mr D. M. Evans, Monsanto Europe SA, Brussels, Belgium.
Mr V. Favoro, Ministry of Agriculture, Fisheries and Forests, Koronivia, Fiji.
Mr T. Fenner, Department of Primary Industry, Konedobu, Papua New Guinea.
Mrs C. A. Ferguson, Bayer UK Ltd, Bury St Edmunds, UK.
Mr I. D. Firman, South Pacific Commission, Noumea, New Caledonia.
Mrs P. Fynn, Shell International Chemical Company Limited, London, UK.
Dr Geh Smee Lan, Malaysian Agricultural Research and Development Institute, Serdang, Malaysia.
Dr I. A. S. Gibson, ODA Liaison Officer, Plant Pathology, Commonwealth Mycological Institute, Kew, UK.
Mr C. P. Haines, Tropical Products Institute, Tropical Stored Products Centre, Slough, UK.
Dr S. D. Haque, Caribbean Agricultural Research and Development Institute, St Augustine, Trinidad.
Dr M. Hollings, Glasshouse Crops Research Institute, Worthing UK.
Dr N. Hrishi, Director, Central Tuber Crops Research Institute, Trivandrum, India.
Dr G. V. H. Jackson, Ministry of Agriculture and Lands, Solomon Islands.
Mr B. Karan, Ministry of Agriculture Fisheries and Forests, Koronivia, Fiji.
Dr M. K. Kamath, Ministry of Agriculture Fisheries and Forests, Nausori, Fiji.
Mr Kaufmann and Mr Huttenbach, Hoechst Aktiengesellschaft, Frankfurt, Germany.
Mr M. Kirby, Ministry of Agriculture Fisheries and Forests, Koronivia, Fiji.
Dr Kradel and Dr Haffke, Crop Protection Division, BASF Aktiengesellschaft, Limburgerhof, Germany.
Mr T. G. La Joubert, Horticultural Research Institute, Pretoria, South Africa.
Dr Lal, Central Tuber Crops Research Institute, Trivandrum, India.
Dr K. Leuschner, International Institute of Tropical Agriculture, Ibadan, Nigeria.
Dr J. Carlos Lozano T., Centro Internacional de Agricultura Tropical, Cali, Colombia.
Mr A. M. MacDonald, Stauffer Chemical S.A., Biggleswade, UK.
Dr R. Meissner, Horticultural Research Institute, Pretoria, South Africa.
Dr F. Monge, Centro Internacional de Agricultura Tropical, Cali, Colombia.
Dr L. A. Mound, British Museum (Natural History), London, UK.
Dr D. Nascimento, Instituto de Pesquisa Agropecuarias do Nordeste, Pernambuco, Brazil.
Mrs D. Nel, Horticultural Research Institute, Pretoria, South Africa.

Dr N. C. Pant, Director Commonwealth Institute of Entomology, London, UK.
Mr C. Parker, ODA Tropical Weeds Specialist, Weed Research Organization, Oxford, UK.
Dr H. C. Passam, Tropical Products Institute, London, UK.
Mr C. H. Perry, Department of Primary Industry, Papua New Guinea.
Dr K. S. Pillai, Central Tuber Crops Research Institute, Trivandrum, India.
Dr D. L. Plucknett, University of Hawaii, Manoa, Hawaii.
Dr R. T. Plumb, Rothamsted Experiment Station, Harpenden, UK.
Dr O. Sampaio, Passos Centro National de Pesquisa de Mandioca e Fruticultura, Cruz das Almas, Brazil.
Dr P. C. Spensley, Director, Tropical Products Institute, London, UK.
Dr J. N. L. Stibick, Department of Primary Industry, Konedobu, Papua New Guinea.
Dr Strydom, Director, Horticultural Research Institute, Pretoria, South Africa.
Dr E. R. Terry, International Institute of Tropical Agriculture, Nigeria.
Dr B. M. Thistleton, Department of Primary Industry, Mount Hagen, Papua New Guinea.
Mr J. Uluinaceva, Ministry of Agriculture Fisheries and Forests, Koronivia, Fiji.
Dr E. V. Wann, USDA, US Vegetable Laboratory, Charleston, USA.
Professor L. A. Wilson, University of the West Indies, St Augustine, Trinidad.
Dr M. Yaseen, Commonwealth Institute of Biological Control, Curepe, Trinidad.

INTRODUCTION

Root crops form an important part of the diet in many areas of the tropics and subtropics. The four main types dealt with in this manual are cultivated, mostly as subsistence crops, in developing countries. Cassava and sweet potatoes are both dicotyledons grown for their edible root tubers, yams and taro (and the other aroids such as eddoe and tannia) are monocotyledons grown for their tubers and corms respectively. All produce starchy storage organs which provide major contributions to the calorific content of the diet, especially in humid tropical countries where many other staple food crops do not grow well. Table 1 shows the estimated annual world production of these four types of root crop.

TABLE 1. PRODUCTION (1000 t) OF ROOT CROPS IN 1975*

Continent	Sweet potato	Cassava	Yams	Taro and other aroids
Africa	5,672	43,972	19,539	3,621
N. and C. America	1,269	769	243	—
S. America	2,729	31,432	48	—
Asia	126,254	28,814	168	619
Oceania	571	221	200	262
Europe	75	—	—	—
World Total	136,570	105,209	20,198	4,502

*From FAO (1976) *Production Year Book* Volume 29.

IPOMOEA BATATAS (L.) POIR. SWEET POTATO

The only economically important member of the family Convolvulaceae is *Ipomoea batatas*. It is not known in the wild state but probably originated in Central America, where it has long been cultivated. It is now grown extensively throughout the tropics and subtropics.

It is a perennial herb with a vine-like habit and great variation in leaf form. Prostrate stems commonly root at the nodes when they contact the soil. Fibrous adventitious roots are produced initially, after which up to ten tubers may develop by secondary thickening (Fig. 1).

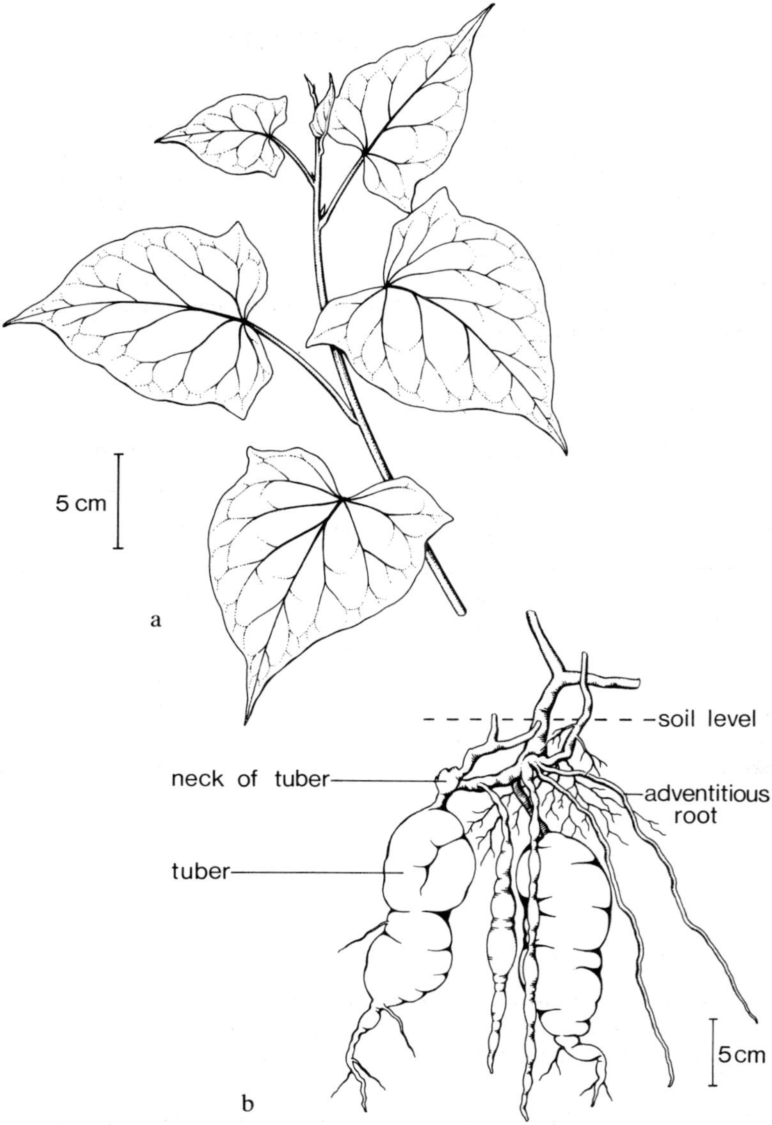

Fig. 1. *Ipomoea batatas,* portion of vine (a) and underground parts (b). (H.R.B.)

Sweet potatoes are grown between 40°N and 32°S but grow best with an average temperature of 24°C, an average rainfall of 75–125 cm and a frost-free growing period of 4–6 months. On the equator they are grown from sea level to about 2,400 m and in warm temperate countries can be grown as a short-term summer crop. In high rainfall areas they are often planted at the end of the rainy season, being relatively drought resistant.

In some areas sweet potatoes are grown as a perennial crop but usually they are regularly harvested and replanted by rooting stem cuttings. Being demanding of nutrients they are usually the first crop grown in a rotation. They are often planted on mounds or ridges. Sweet potatoes do not store well unless first cured at 30–32° and 85–90% r.h. for 1–2 weeks.

MANIHOT ESCULENTA CRANTZ CASSAVA, MANDIOCA, TAPIOCA

M. esculenta is a member of the Euphorbiaceae. It is not known in the wild state but has long been cultivated in Central America and Brazil and was more recently carried to Africa (where it is now an important subsistence crop) Asia and the Pacific Islands.

Cassava is essentially a crop of the lowland tropics but will grow up to 1,500 m near the equator. It requires an annual rainfall of at least 50 cm or preferably more, but can tolerate periods of drought. It grows best on light fertile soils but can grow on almost any type of soil provided it is well drained. An economic crop can be obtained even under marginal conditions or on exhausted soils. Cassava is often used as the last crop on land before it is returned to fallow.

It is a small shrub up to about 5 m high which produces a number of tubers by secondary thickening of some of its fibrous roots (Fig. 2). Usually 5–15 tubers between 15 and 100 cm long are produced. Many cultivars are recognised, which can be roughly divided into two main groups; sweet cassava and bitter cassava. The tubers of all cultivars contain the highly toxic hydrogen cyanide but in the former the concentration is much lower and it is largely confined to the rind, while in the latter it is distributed at higher levels throughout the tuber and must be eliminated by boiling, drying, roasting or fermenting before use.

Cassava is often interplanted with other crops. Hardwood stem cuttings, usually about 300 mm long are planted either on mounds or ridges or on the flat. Cultivation of the soil improves the yield but after planting the crop requires very little attention though early weeding is beneficial. Tubers do not store well, and are usually harvested as required.

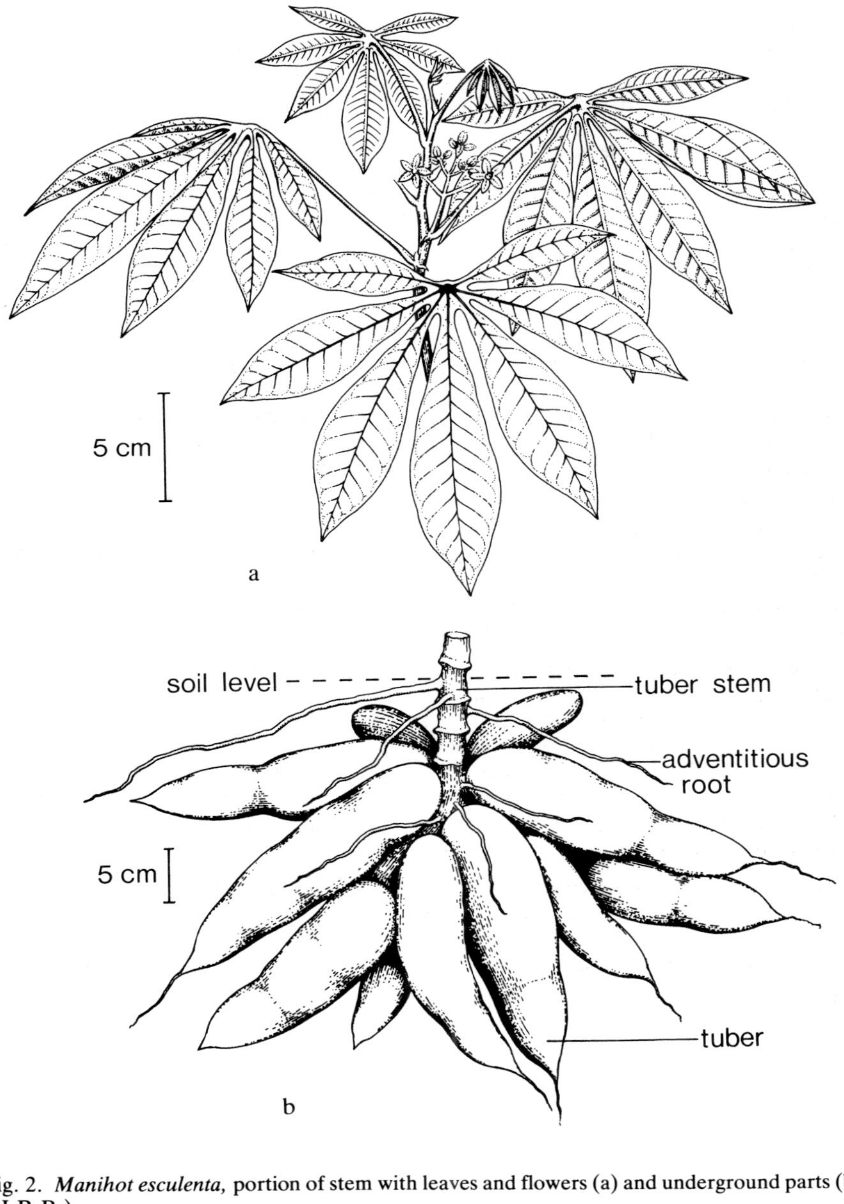

Fig. 2. *Manihot esculenta,* portion of stem with leaves and flowers (a) and underground parts (b). (H.R.B.)

DIOSCOREA SPP. YAMS

Yams belong to the genus *Dioscorea,* the main genus of the family Dioscoreaceae, which is widespread throughout the tropics and subtropics with over 600 species. Only a few produce edible tubers of which *D. alata, D. rotundata, D. cayenensis* and *D. esculenta* are the most important.

D. alata L., the greater yam was first cultivated in S.E. Asia and is now unknown in the wild state. It is widely grown in Asia and the Pacific, is less favoured in Africa, but much grown in the Caribbean. Usually a single tuber is produced, there being considerable variation in shape, size and colour between the large number of cultivars which are known (Figs. 3a and b).

D. rotundata Poir., the white or white Guinea yam, is the most important species in W. Africa, where it originated. Tubers are usually cylindrical but size, shape and colour vary according to cultivar, of which there are a great many. It is believed to be closely related to *D. cayenensis.*

D. cayenensis Lam., the yellow Guinea yam is a native of W. Africa, where it occurs widely and is extensively cultivated. Together with *D. rotundata* it is also grown in the Caribbean. It is hardier than *D. rotundata* and can be harvested over a longer period, although the flavour is not as good.

D. esculenta (Lour) Burk., the lesser yam, originated in Indo-China and is extensively grown in S.E. Asia and elsewhere in the tropics, although it is still rare in Africa. It produces large numbers of small ovoid tubers resembling sweet potatoes in appearance. It is high yielding and is well adapted to mechanical harvesting.

D. trifida L., the cush-cush yam, is the only cultivated species native to S. America and is much valued in the Caribbean for its flavour. It bears a number of small tubers (Figs. 3c and d).

Other cultivated species include *D. bulbifera* L., the potato yam, which is widespread throughout Africa and Asia. The subterranean parts are much reduced but edible aerial bulbils are produced. *D. opposita* Thunb., the Chinese yam, is cultivated in China and S.E. Asia and produces very long spindle-shaped tubers. *D. dumetorum* (Kunth) Pax. the African bitter yam and *D. hispida* Dennst., the Asiatic bitter yam are toxic, but the poison can be removed and they are collected and eaten in times of food shortage. Several other species including *D. pentaphylla* L. and *D. nummularia* Lam. are of lesser importance as food plants.

The major yam species are tropical plants and do not grow well below 20°C. They require at least 100 cm rainfall a year. Forest species, for example *D. cayenensis,* grow almost continuously and cannot tolerate more than 2–3 months without rain; savanna species such as *D. alata* and *D. rotundata* require a growing period of 7–8 months.

Yams are perennial plants with a vine-like habit. According to species they produce one large tuber or several smaller ones. They are normally grown as an annual crop, seed tubers or pieces of tuber being planted, usually on mounds or ridges, or sometimes in trenches or holes filled with organic matter. Stakes or strings are provided as support for the aerial parts, or trees may be left after clearing for this purpose. They are not well suited to mechanical cultivation and are mostly grown as a subsistence crop, but have a high labour requirement. In shifting cultivation or rotations they are the first crop

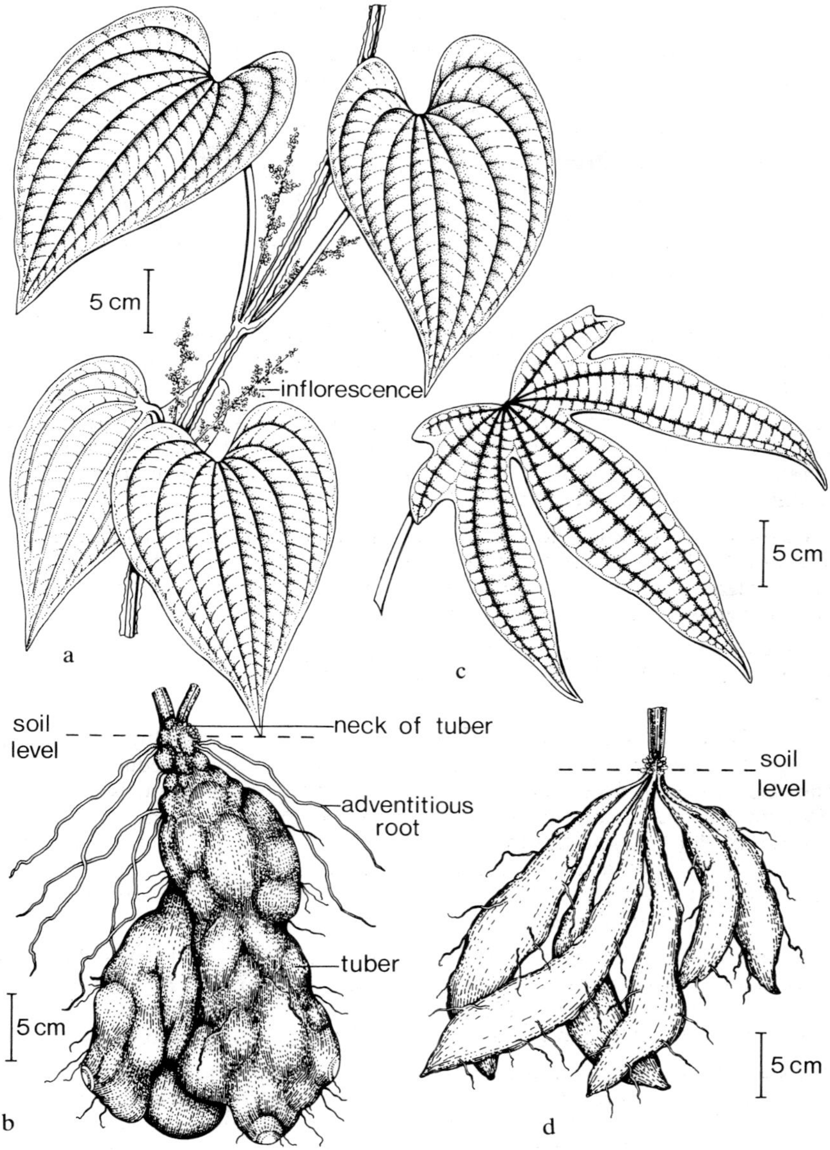

Fig. 3. *Dioscorea alata,* portion of vine showing leaves and inflorescence (a) and tuber (b). *D. trifida* leaf (c) and tubers (d). (H.R.B.)

planted after clearing as they are demanding of plant nutrients. In many parts of Africa yams are being displaced to some extent by sweet potatoes or cassava, which require less labour and give greater yields, especially on degraded or marginal soils.

COLOCASIA ESCULENTA (L.) SCHOTT TARO, DASHEEN, COCOYAM

Taro is a member of the family Araceae, and occurs wild in S.E. Asia. It is cultivated throughout the humid tropics and is of great importance in the Pacific Islands. Taro requires plentiful rainfall and heavy fertile soils: it is sometimes grown under flood irrigation. It is a herb 1–2 m tall with a starchy cylindrical corm up to 30 cm long and 15 cm diameter, which bears a whorl of large shield-shaped leaves on erect petioles (Fig. 4a).

The plants are propagated by tops of mature corms or by side tubers or suckers. The crop is normally grown in flat, wet areas.

Eddoe, although regarded by some authorities as a distinct species, *Colocasia antiquorum,* is generally accepted as a subspecies *C. esculenta* ssp. *antiquorum.* It is grown in the W. Indies, and many W. African forms of cocoyam are similar to eddoe. It is hardier than taro and can be grown with less rainfall and on lighter, poorer soils. It is generally similar to taro except that large numbers of cormels are produced around the main corm.

XANTHOSOMA SAGITTIFOLIUM (L.) SCHOTT TANNIA, YAUTIA OCUMO

X. sagittifolium is also known in W. Africa as cocoyam. It originated in tropical America and the W. Indies and in only recent times has spread to other areas of the tropics. It requires high rainfall and soil moisture and is usually grown on ridges. The corm bears ten or more lateral tubers or cormels 15–25 cm long (Fig. 4b). It is hardier and often higher yielding than *Colocasia,* which it is displacing in some areas.

Several other aroids, including *Alocasia, Amorphophallus* and *Cyrtosperma* spp. are of minor, local importance.

The edible aroids are attacked by many of the same pests and diseases and will be dealt with under taro.

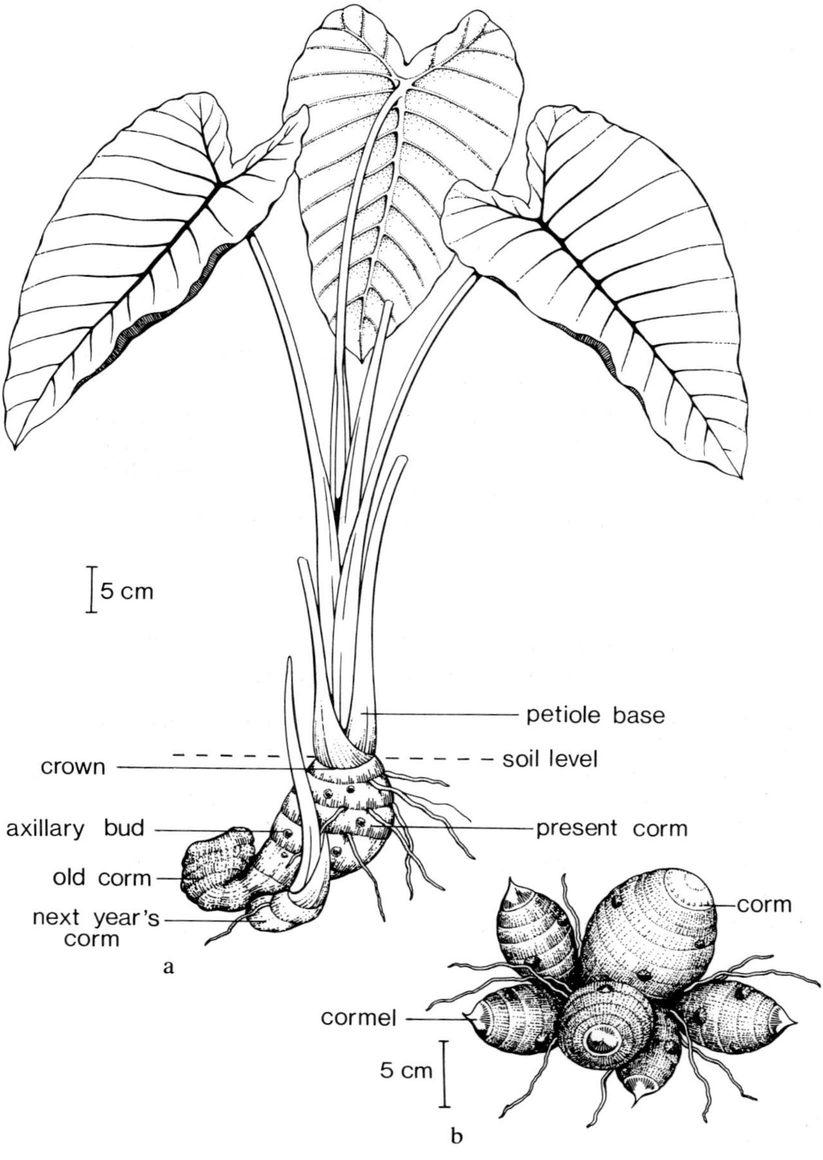

Fig. 4. *Colocasia esculenta,* entire plant (a) and corm of *Xanthosoma sagittifolium* with cormels (b). (H.R.B.)

Bibliography

ACLAND, J. D. (1971). *East African crops.* pp. 252. Longman, London for FAO.

COBLEY, L. S. (1976). *Introduction to the botany of tropical crops.* 2nd Edition, revised by W. M. Steele. pp. 371. Longman, London.

COCK, J., MACINTYRE, R. and GRAHAM, M. (Eds) (1977). *Proceedings of the Fourth Symposium of the International Society for Tropical Root Crops,* Cali, Colombia, 1976. pp. 277. International Development Research Centre, Canada.

COURSEY, D. G. (1967). *Yams.* pp. 230. Longman, London.

COURSEY, D. G. (1968). The edible aroids. *World Crops* 20(4): 25–30.

COURSEY, D. G. and HAYNES, P. H. (1970). Root crops and their potential as food in the tropics. *World Crops* 22: 261–265.

IRVINE, F. R. (1969). *West African crops.* pp. 272. Oxford University Press.

KAY, D. E. (1974). *Root crops. Crop and product digest* No. 2. pp. 245. Tropical Products Institute, London.

NATIONAL ACADEMY OF SCIENCES. (1975). *Underexploited plants of promising economic value.* pp. 189. National Academy of Sciences, Washington.

PURSEGLOVE, J. W. (1968). *Tropical crops; Dicotyledons* (2 volumes) pp. 717, *Monocotyledons* (2 volumes) pp. 607. Longman, London.

WEED CONTROL IN TROPICAL ROOT CROPS*

It is difficult to generalise about weed control in tropical and subtropical root crops due to the widely varying standards of husbandry under which they are raised. For example, cassava may be grown as a minimum input subsistence crop on small plots on land about to be abandoned and possibly interplanted with other crops or as a well managed semi-plantation monocrop to produce starch for industry. Thus one may find recommendations for sophisticated programmes of weed control that are quite inappropriate (and not intended) for peasant agriculture. Indeed before one can begin sensibly to consider the degree and type of weed control to be aimed at in any particular situation it is necessary to have a clear picture of its economics. For example, in the Solomon Islands, where sweet potato and taro are the staple crops and where land is readily available, it is more sensible to raise vigorous cultivars that require little weeding than cultivars that give a higher yield per unit area but which are more demanding. Clearly, where land is at a premium the same criteria are unlikely to apply. Perhaps now that the constantly increasing cost of agricultural chemicals is bringing into question the tendency to emphasise high-yielding cultivars that demand rigorous standards of cultivation (fertilizers, pest, disease and weed control) plant breeders will place a cultivar's ability to compete with weeds on a par with its resistance to pests and diseases. Perhaps one reason this has not been done hitherto is because it is not generally realised how much crop yields may be reduced by weeds—a view resulting, no doubt, from weeds always having been with us and because their effects are much less obvious than those of pests and diseases.

It is general knowledge that weeds decrease crop yields by competing for water, nutrients and light and sometimes by acting as hosts for pests and diseases; less appreciated is the phenomenon of allelopathy (the excretion or secretion by certain plants of toxins that adversely affect others) which may be quite widespread. Although, under certain circumstances weeds can be beneficial e.g. by recirculating nutrients through the soil profile (particularly in the case of species that have the ability to concentrate scarce elements that they release on their death and decay), the protection they afford the soil against erosion and their ability to act as hosts for the predators of crop pests and diseases there is no doubt that, on balance, weeds are nearly always deleterious. Competition is particularly damaging during the early life of a crop except for a brief period directly after planting—probably of about 1–2 weeks in the case of sweet potato and 3–4 weeks for yams, cassava and taro. Studies in South America, Africa and the Caribbean have shown that the period of the greatest susceptibility to weed competition lasts for about the first quarter to one third of a crop's life. The length of this period is influenced to some extent by the climate, weed species, crop variety, fertilizer and water regimes and also by factors such as planting density (which latter has rarely been taken into account in competition studies).

---

*This chapter was written by Mr L. Kasasian.

Cultural weed control

It is clear that as clean a seedbed as possible is desirable to assist the crop to establish itself. In slash and burn agriculture this may be achieved by planting as soon after the burn as possible, while on previously cultivated land good primary cultivation is important. In some cases it can be useful to apply herbicides before or during land preparation to eradicate, or at least to reduce to manageable proportions, particularly noxious weeds.

Manual weeding is still the most common form of weeding in tropical root crops and usually accounts for about one third of the total labour used in raising these crops. Because the work is so laborious there is an almost universal tendency to leave the first weeding after planting until too late, by which time the weeds will have done irreparable damage. Once it is realised that early weeding is vital and that, moreover, it is far easier and quicker to deal with weeds when they are small, then it may be accepted that (if soil conditions are suitable) light Dutch hoes can save much time and effort. The tendency has so far been to reject these tools mainly because they will not deal with established weeds.

Chemical weed control

In many situations herbicides can be extremely useful, not only by making weeding much easier and quicker but also by making it more effective (e.g. by making it possible to keep the young crop really weed-free), and in avoiding the mechanical damage to the crop's roots and shoots that may occur during post-planting cultivations. Herbicides may also increase profitability by making it possible to farm more land.

Only recently have different methods of weed control come to be regarded as complementary, rather than as alternatives, to each other. This concept of 'integrated control' may mean no more than the use of herbicides to eradicate deep-rooted perennial weeds or other weeds that it is impracticable to kill in any other way and the use of mechanical methods for general weed control. It may mean the use of chemicals in only certain crops in a rotation (e.g. if labour is short after they are planted or if herbicides are particularly effective in that crop); or using weedkillers on land difficult to weed in other ways because it is stony, very wet or very dry, or simply as a temporary measure to reduce weeds to a manageable level. A major problem in using herbicides (quite apart from their cost and availability) is the need for timely and accurate application if they are to work safely and efficiently. Where, as often happens, herbicides are spectacularly effective at first there is a natural tendency to disregard the few weeds they fail to kill—this is a serious mistake, the result of which is to select out a more resistant weed flora. The new flora will inevitably proliferate and may be more harmful and difficult to control than that originally present, requiring more sophisticated and expensive chemicals for control. However, good, disinterested extension work should prevent this problem arising. One way to avoid it is to use mixtures of different herbicides. Another method is to alternate the herbicides used. This is of course easier when crop rotation is being practised, although one must then be careful to choose herbicides that will not leave residues in the soil to affect the following crop. This is rarely a problem with yams, sweet potatoes and taro partly as they are in the ground long enough for most herbicides to have dissipated before the next crop is planted and partly because they are normally grown under conditions where herbicide persistence is not high. Residues may, however, be a problem after cassava when it is grown under dry conditions where degradation is slow. The common practice of inter-planting other crops with root crops makes it more difficult to use herbicides as all the crops are unlikely to tolerate a given herbicide equally well. This is possible in some cases, however, such as in Barbados where yams and sugarcane are intercropped when atrazine may be used on both or where contact herbicides are applied with a shielded nozzle.

As tropical root crops tend to form the major part of the diet of their consumers, it is especially important to ensure that no harmful residues of agricultural chemicals are to be found in the crop at harvest and that the nutrient quality of the crop is not adversely affected. These factors need to be balanced against the greater total food production that can result from efficient weed control.

Types of herbicides

Chemicals may be applied to standing vegetation before planting either to make subsequent cultivations easier or even as an alternative to them. Glyphosate and paraquat are particularly useful for this purpose as neither compound leaves any active residues in the soil to affect the new crop. Glyphosate is a systemic herbicide and is, therefore, especially useful for perennial species. A major drawback in the wet tropics, however, is that several hours of dry weather after spraying are necessary for good results. Glyphosate kills many grasses and broadleaved weeds and is very effective on sedges. Paraquat is not systemic and is, therefore, most useful for annual weeds, although its desiccant action can be useful in making almost any vegetation more combustible. Paraquat is unaffected by rain even almost immediately after spraying.

After seedbed preparation, but before planting, chemicals may be applied to prevent the growth of weeds from seeds and roots in the soil. Compounds applied at this time include trifluralin, dinitramine, butylate, vernolate and EPTC; all these compounds are volatile and must be mixed into the top 5–10 cm of the soil shortly after application and therefore demand both a very good seedbed and suitable equipment to effect incorporation (a powered rotary cultivator is ideal). However, in irrigated agriculture EPTC and vernolate may be applied through the irrigation water (herbigation, see below) instead of by spraying and incorporating and then the seedbed is not so critical. Pre-planting treatments are affected relatively little by soil moisture.

Pre-emergence treatments (i.e. those applied after planting a crop but before it, or the weeds, emerge) act by forming a layer of herbicide on the soil that is effective mainly in preventing weed infestation from seed in the soil. These treatments are intended primarily to keep the land weed-free until the crop has become established. Once the treated layer of soil is disturbed (by cultivations etc) the treatments lose their effectiveness. A moist soil or rain or irrigation within a week or so of application is generally necessary for satisfactory results.

Post-emergence treatments (applied after crop and weed emergence) may be either contact or systemic sprays. Contact sprays (e.g. paraquat) kill the tender shoots of any plants with which they come into contact, but not normally the roots. The effect of these 'chemical hoes' is therefore short-lived, but it may be prolonged by the addition of a soil-acting pre-emergence herbicide. As contact sprays are general killers they need to be directed away from tender crop growth and a spray shield may be necessary (Fig. 5). Systemic sprays kill both the shoots and roots of susceptible plants and are therefore especially useful for established perennial weeds. These chemicals generally kill either only grasses (e.g. dalapon) or only broadleaved weeds (e.g. 2,4-D); however, some, like glyphosate, are active on both groups of plants. Directed application of systemic sprays is necessary unless the crop is tolerant, but even normally tolerant crops may be injured through root uptake in certain circumstances.

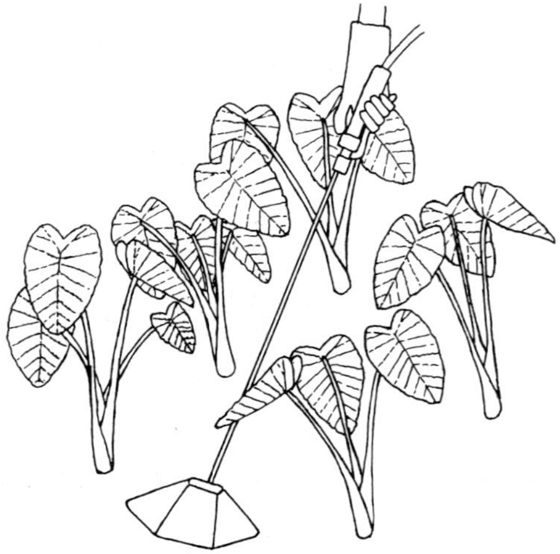

Fig. 5. A spray shield attached to a lance from a knapsack sprayer. (COPR)

Application

Herbicides are usually applied as sprays when it is advisable to use low pressure nozzles to decrease the risk of spray drift. Even and rapid application of overall sprays is most easily made by using a boom with several jets; large booms are tractor-mounted or trailed, but small light-weight booms are available for use with knapsack sprayers. Single nozzles can give acceptable results for overall spraying but are generally more suitable for directed spraying.

The volume rate (the number of litres/hectare or gallons/acre) is of minor importance so long as good coverage is achieved and wasteful run-off of foliar sprays is avoided. However, the volume rate must be determined so as to know how much water to mix the herbicide with. This initial calibration is best carried out by pouring a known volume of water into the sprayer and then spraying a measured area of ground exactly as if one were spraying the herbicide; one can then easily calculate how much water has been used by subtracting the amount left in the sprayer from that originally added and thus calculate the volume rate. For example: pour 10 *l* of water into the knapsack sprayer, mark out an area of land, e.g. 10 × 10 m, and spray—if after spraying 6 *l* of water were left then 4 *l* was used for the 100 m^2 sprayed and therefore the volume rate per hectare would be $\frac{4 \times 10000}{100} = 400\ l$. Then if one wanted to apply 3 kg/ha a.i. of 80% atrazine one would mix 3.75 kg (i.e. $\frac{3 \times 100}{80}$) with each 400 *l* of water*. Recalibration is necessary periodically and whenever the nozzle, sprayer or speed of working is changed (due to change in terrain, for example).

'Herbigation' is the application of a herbicide through the irrigation water. With flood irrigation this is done by dripping the chemical into the water entering the field using a constant-flow device (see Figs. 6 and 7). Stones are placed in the water under the drip tube to introduce turbulence and thus effectively mix the chemical with the water. Herbigation may also be carried out with sprinkler irrigation, but this requires the chemical to be fed into the spray line under pressure.

Fig. 6. Constant flow device for metering herbicide into irrigation water. (Staufer Chemical Company)

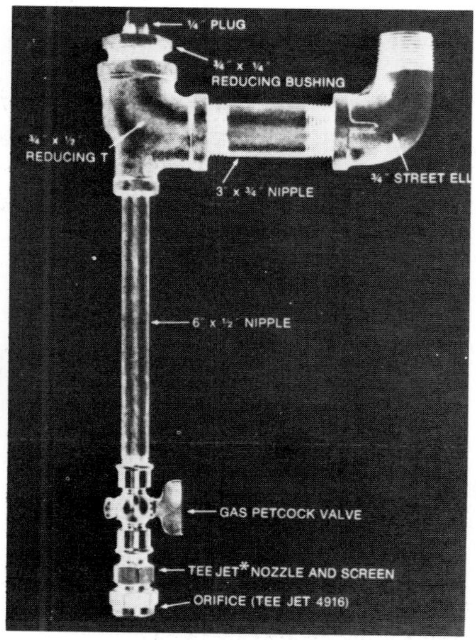

Fig. 7. Close up of constant flow device. *Trade mark of Spraying Systems Co. (Staufer Chemical Company)

*If the spray tank had a capacity of (say) 15 *l* then each tankful would contain approximately 140g of herbicide (i.e. $3.75 \times \frac{15}{400}$ kg).

TABLE 2. THE MOST PROMISING HERBICIDES FOR USE IN TROPICAL ROOT CROPS

Sweet potato	Cassava	Yams	Taro
		Pre-Planting	
EPTC Nitralin Trifluralin Vernolate	Butylate Trifluralin		Trifluralin
		Pre-emergence	
Alachlor Chloramben Chlorbromuron Chlorthal Diphenamid Metobromuron Napropamide Noruron Prometryne	Alachlor Diuron Fluometuron TCA	Alachlor Ametryne Atrazine Chlorthal Linuron Metribuzin Simazine TCA	Ametryne Atrazine Chloramben Linuron Methazole Monolinuron Nitrofen Oxadiazon Prometryne Simazine
		Post-emergence	
Paraquat	Dalapon* Diuron Glyphosate* MSMA* Paraquat*	Dalapon Paraquat	Dalapon Paraquat Propanil

*Contact between these herbicides and cassava foliage can be harmful under some circumstances and should be avoided.

Some herbicides may be applied as granules, but generally these are only satisfactory in aquatic situations, e.g. paddy taro and rice. Granular herbicides may be distributed by fertilizer spreaders, seeders, by tins with suitable sized holes in the bottom or simply by hand, in which case a glove should be worn.

Precautions in the use of herbicides

There is no such thing as a completely safe pesticide, although many are safer than a number of other chemicals in common use. Always observe the following precautions:

1. Read the label on the container carefully and completely.
2. Keep containers and mixed sprays away from children, animals, foodstuffs and fertilizers. Wash out used containers and dispose of them and unused sprays safely.
3. Never keep pesticides in unlabelled containers.
4. Never mix chemicals by hand, always use a special stirrer.

5. Never try to clear a blocked nozzle by sucking or blowing—use a sharpened match or piece of wood.
6. If chemicals get in the eyes, flush with plenty of clean water immediately; if on the skin, wash with soap and plenty of water immediately.
7. Always wash before eating or smoking.
8. Wear adequate protective clothing, at least long trousers and shoes, and wash clothing frequently, preferably daily.
9. If unwell after spraying consult a doctor without delay.
10. Avoid contaminating waterways.
11. Do not endanger wildlife during spraying or expose them to toxic residues.

Sweet potato

Competition studies in Trinidad showed the crop to be most susceptible to weed competition during the first third of its life, which meant that weeding was necessary for at least five weeks. However, where vigorous cultivars are grown such intensive weeding may be unnecessary and in some areas the crop is so competitive that it is planted directly after land clearance as a cleaning crop. In Australia, where sweet potatoes are planted on well-prepared land, only a single weeding is normally required and a multiple-row finger cultivator, used before the vines start to run, is regarded as being ideal.

As sweet potatoes are planted as cuttings bearing expanded green leaves the crop tends to be less tolerant of overall sprays than the other root crops. Recent experiments in the southern USA have shown that no herbicide treatment, with or without cultivations, yielded as well as handweeding. Moreover, the ridging usually carried out after establishment reduces the effectiveness of any residual herbicides previously used. However, herbicides can be useful especially during establishment.

Of the incorporated pre-planting treatments trifluralin and nitralin are excellent for the control of grasses and some broadleaved weeds growing from seed, while EPTC and vernolate (which control rather fewer grass and broadleaved species) are effective on sedges (including *Cyperus* spp.).

After planting, the most widely used treatments appear to be diphenamid for grass control and chloramben and chlorthal, which are better on broadleaved weeds. All three may be applied as overall treatments and mixtures of diphenamid with either chloramben or chlorthal are frequently used. Chlorthal has been found to reduce the formation of adventitious roots on runners and thus helps in avoiding the production of numerous very small tubers. Chlorthal, like EPTC and vernolate, may be applied by herbigation, although this method of application is unlikely to be satisfactory if the crop is grown on ridges. There are a number of reports in the literature indicating that diphenamid, chlorthal, chloramben and EPTC have no adverse effect on the nutrient or storage quality of sweet potatoes. More recently alachlor has given excellent results on grasses; it is used at 0.5–1 kg a.i./ha in Malaysia and at 1.5–3 kg a.i./ha in New Zealand in sweet potatoes.

Of the more active herbicides promise has been shown by overall applications of chlorbromuron in the USA, West Africa and the Caribbean and also by noruron, metobromuron and prometryne in the Caribbean, while in Japan napropamide at 3–4 kg a.i./ha and granular prometryne have given good results. Where it is normal practice to turn the vines to prevent the runners rooting, it is possible to make a directed application of paraquat (using a shielded nozzle) at this time (Fig. 8). In Papua New Guinea the local cultivars are said to tolerate overall applications of paraquat.

Fig. 8. Inter-row weeding using paraquat in sweet potatoes. (L. Kasasian)

Cassava

An extensive programme on the control and effects of weeds in cassava has been carried out at CIAT in Colombia. In a survey of nearly 300 farms in five cassava-growing regions it was found that manual weeding accounted for about 50% of the total labour requirement of the crop (and approximately one third of the total cost of production). However, in Nigeria, where weed competition appears to be less serious, weeding accounts for only about 25% of the labour used. In Colombia it was necessary to keep the crop weed-free for at least four months to secure the maximum yield, whereas in Nigeria only 2–3 months is said to be necessary—this may be no more than a reflection of the relative speed of growth of the crop in South America and West Africa.

Studies at CIAT on the effects of plant density on the competitive ability of the crop have shown that where only 1–2 weedings were carried out the best yields were obtained with densities of 15,000–20,000 plants/ha of the short non-branching cultivar Mexico 11 and 20,000–25,000 plants/ha of the taller branching cultivar CMC-39, whereas where weeding was better both varieties did better at a lower density of 15,000 plants/ha. Other experiments at CIAT showed that under weedfree conditions a much lower density (6000–8000 plants/ha) gave the highest yields.

In a comparison of chemical and manual weeding at CIAT it was found that three handweedings outyielded all other treatments; however, a pre-emergence application of diuron followed by a single handweeding was nearly as good. Pre-planting incorporated treatments that give particularly good results in cassava are trifluralin (Fig. 9) at 0.25–1.0 kg a.i./ha (for grasses and some broadleaved weeds growing from seed) and butylate at 2–4 kg a.i./ha (where sedges are also a problem).

Fig. 9. Result of field trial using trifluralin in cassava in Colombia. (J. D. Doll)

Many substituted urea and triazine herbicides have given good early general weed control when applied pre-emergence and the most consistent results seem to have been obtained with diuron and fluometuron at 1.5–3.0 kg a.i./ha. Alachlor at 1–2 kg a.i./ha, or occasionally TCA at 5–10 kg a.i./ha, may be added to either of these to improve grass control (Fig. 10) although there is some evidence that both alachlor and TCA increase the drought susceptibility of certain crops. It is likely that very good general weed control would be obtained by pre-planting applications of trifluralin followed by pre-emergence application of diuron or fluometuron.

Fig. 10. Result of field trial using fluometuron + alachlor in cassava in Colombia. (J. D. Doll)

Once the crop is established basally-directed applications of paraquat (alone or mixed with diuron) will give good general weed control (Figs. 11 and 12); shielded application is necessary until the cassava stems have hardened and turned brown. At this time grass weeds may be controlled by directed sprays of dalapon and MSMA without the use of a spray shield. Glyphosate can be very useful, but shielded application is essential even in a well established crop. All of these herbicides can under some circumstances severely injure cassava if they come into contact with the foliage.

Yams

In both West Africa and the Caribbean weed control was found to be critical during the first three months after planting edible yams. Steroid-producing yams, raised for medicinal purposes, are generally not harvested until the crop is 3–4 years old and it is interesting that in these much longer-term crops it is still during the first quarter to one third of the crop's life that weeds are most harmful. There are reports that continuous weeding from planting till harvest may be injurious by increasing the rotting of tubers (perhaps as a result of higher soil temperatures and moisture in the absence of weeds).

Burying weeds in the ridges or mounds on which yams are usually planted gives a clean seedbed suitable for the application of pre-emergence herbicides. In both the Caribbean and West Africa a mixture of atrazine 3 kg a.i./ha and TCA 5 kg a.i./ha (the

Fig. 11. Application of paraquat to cassava. The lance is held to give a horizontal spray pattern to avoid leaf scorch. (R. S. Harper (1974), *PANS* 20(2): 185–189)

Fig. 12. Experimental plot of cassava sprayed with paraquat at 14-day intervals. (R. S. Harper, (1974), *PANS* 20(2): 185–189)

latter to improve grass control) has given very good results; however, the more recently introduced alachlor should have a longer residual effect than TCA. Alachlor is recommended by the manufacturers for use in yams at 2 kg a.i./ha. In West Africa a mixture of chlorthal 5 kg a.i./ha + linuron 1.5 kg a.i./ha has also proved to be promising.

At one time diuron was recommended in the West Indies, but subsequently it was found to cause injury on occasions. Initially diuron also gave good results in Brazil, but later simazine was found to be better. Most recently metribuzin and ametryne have shown promise in yams. In the established crop dalapon at 5 kg a.i./ha has been successfully used for grass control (applied by shielded nozzle), and paraquat may be similarly applied.

Taro

The weed problems and herbicides that can be used are quite different in paddy and upland taro. In Hawaii, a number of chemicals have given good results in paddy taro when applied by herbigation, providing the herbicides were mixed very thoroughly with the irrigation water. Herbicides successfully applied in this way are nitrofen at 3–6 kg a.i./ha, chloramben 3 kg a.i./ha, methazole 3 kg a.i./ha and oxadiazon 1.5–3 kg a.i./ha. Nitrofen, which is registered for use in taro in Hawaii, may also be applied as a spray after draining, with a second application when the fields are redrained for fertilizer application. Spraying may cause some leaf scorch, however, which is avoided by herbigation. It seems likely that the granular formulations of nitrofen (developed for use in paddy rice) might also be useful.

At one time propanil was used for grass control in Hawaii, but only early applications (before taro had three leaves) were safe, a degree of water control permitting draining before spraying and re-flooding 24 hours later was necessary and only very young grass weeds were susceptible.

In the upland crop good results have been given by ametryne and prometryne in the Caribbean and the Solomon Islands and by linuron, monolinuron, atrazine and simazine in Fiji. These compounds are all applied as overall sprays directly after planting and before any crop leaves open. In Fiji good results have also been given by directed applications of paraquat, while in Samoa low rates of dalapon (3 kg/ha) have been successfully used after planting.

TABLE 3. MAIN USES OF HERBICIDES MENTIONED IN THE TEXT

Common name	Usual rate (kg a.i./ha)	Main uses	
		Pre-planting	
Butylate	2–4	Control many grasses from seed, some broadleaved weeds and some sedges	All need incorporating (e.g. by use of a powered rotary cultivator)
EPTC	2–6		
Vernolate	2–6		
Trifluralin	0.25–1	Controls many grasses and some broadleaved weeds from seed	
		Pre-emergence	
Alachlor	1—3	For control of grasses, especially growing from seed. Usually mixed with one of the herbicides below	
Diphenamid	4–8		
TCA	5–10		
Chloramben	1–4	Generally more effective on broadleaved weeds growing from seed. Of short to moderate persistence	
Chlorthal	4–12		
Methazole	1–4		
Nitrofen	1–4		
Ametryne	1–3	Mainly for control of weeds growing from seed, especially broadleaved weeds. Most have good persistence	
Atrazine	1–3		
Chlorbromuron	1–3		
Diuron	1–3		
Fluometuron	1–3		
Linuron	1–3		
Metribuzin	0.5–1.5		
Noruron	1–3		
Oxadiazon	1–4		
Prometryne	1–3		
Simazine	1–3		

Common name	Usual rate (kg a.i./ha)	Main uses
		Post-emergence
Dalapon	3–10	Control many emerged annual and perennial grasses
MSMA	2—6	
Glyphosate	1–4	Non-selective broad spectrum control
Paraquat	0.6	General contact killer
Propanil	2–6	Mainly for control of *Echinochloa* spp.

Bibliography

CONSTANTIN, R. J., HERNANDEZ, T. P. and WASCOM, B. W. (1975). Effect of herbicides on quality of sweet potatoes. *Proceedings Southern Weed Science Society* 28: 194–200.

DE LA PENA, R. S, and PLUCKNETT, D. L. (1971). Application of herbicides through irrigation water for weed control in lowland taro (*Colocasia esculenta*). *Proceedings 3rd Conference Asian-Pacific Weed Science Society.*

DOLL, J. D. and PIEDRAHITA W. C. (1976). *Methods of weed control in cassava.* pp. 21. CIAT Bulletin Series EE-21. Cali, Colombia.

DOLL, J., PER PINSTRUPANDERSEN and DIAZ, R. (1977). An agro-economic survey of the weeds and weeding practices in cassava (*Manihot esculenta* Crantz) in Colombia. *Weed Research* 17: 153–160.

KASASIAN, L. (1971). *Weed control in the tropics.* pp. 307. Leonard Hill, London.

MOODY, K. and EZUMAH, H. C. (1974). Weed control in major tropical root and tuber crops—a review. *PANS* 20: 292–299.

ONOCHIE, B. E. (1975). Critical periods for weed control in cassava in Nigeria. *PANS* 21: 54–57.

RODENTS*

The extent of the problem

Wherever cassava, yams, taro or sweet potatoes are grown there is a chance they will be attacked and eaten by rodents. The damage may be seen immediately when the roots are eaten into at ground level as sometimes happens with taro (Figs. 13–15), or may not be apparent until the crop is harvested if rats have burrowed and eaten away the roots from beneath. Little attention has been paid to rodent damage of tropical root crops, probably because they are usually grown as subsistence foods and not on a commercial scale. Plots are often small and the damage that occurs, while sometimes of considerable local importance, may be difficult to investigate because of the remoteness of the plots.

Fig. 13. Taro eaten by rats. (Pest Infestation Control Laboratory)

Fig. 14. Close up of rat damage to a taro plant. (Pest Infestation Control Laboratory)

A worldwide survey of rodent damage has indicated very variable assessments of damage to root crops. Losses of the order of 5–15% in West and Central Africa and up to 50–60% in the Cayman Islands in the West Indies were reported. Although losses are also indicated from Sri Lanka, Sarawak, Japan and western Turkey, no quantitative estimates of damage are given.

---

*This chapter was written by Miss E. J. Taylor, Pest Infestation Control Laboratory, Tolworth, UK.

Fig. 15. Taro, showing rat eaten core.
(Pest Infestation Control Laboratory)

Rather more is known about the situation in the oceanic Pacific Islands. Here root crops form an important part of the diet of the people and are widely grown in small plots and gardens. Estimates of 15–20% are given for the Society Islands and up to 30% for the Solomon Islands. Research at the Dala Experimental Station, Malaita, has shown that rat attacks on sweet potatoes begin when the crop is about 100 days old and also that rats showed a preference for two cultivars out of 25 during a trial of different varieties. A simple method of measuring damage to sweet potatoes involves weighing the tubers at harvest and calculating the edible fraction of the rat-damaged ones.

Rodent attack may continue after harvest in farm and village stores, but there is little or no information on the extent of these losses.

Species that cause damage in the field

The species that damage growing root crops are usually the common pest rodents of other crops as well. The relatively few cases where the actual species responsible has been identified, are shown in Table 4.

Species that cause damage in stores

The species responsible for damage in stores depend on where the store is situated. Stores in small rural villages or on farms may well be entered by the same rodents that attack the growing crops. In the Indian sub-continent these will include *Bandicota bengalensis*; in Africa, *Mastomys natalensis*; and in the Far East and South Pacific region, *Rattus exulans*.

TABLE 4. SOME RODENTS REPORTED TO ATTACK GROWING ROOT CROPS*

Region	Scientific and common names	Remarks
West Africa	*Cricetomys gambianus* giant Gambian rat; *Arvicanthis niloticus* harsh-furred Nile rat; *Thryonomys swinderianus* cutting-grass	Information from Nigeria only. *Thryonomys* is eaten by the people who are thus reluctant to use poisons against it
Central Africa	—	Inadequate information.
Southern Africa	*Pedetes capensis* and *Pedetes surdaster* spring hass, ntoile, tshipo, inziponide, goob	Eat fleshy roots as well as other crops. Controlled by hunting and prized as food (Walker, 1975)
East Africa	*Hystrix* sp. porcupine	Stated to be a pest of taro
	Tachyoryctes spp. mole-rats	Pests on farmlands, feeding on, among other things, tubers, rhizomes, bulbs and corms. Sweet potato tubers were found in burrow systems (Jarvis and Sale, 1971)
Mediterranean and Middle East	—	Indequate information
South Asia	Species not identified	In Sri Lanka yams and other tubers are attacked and the tubers and rhizomes eaten, but the losses are not assessed. Baiting with anticoagulants is successful when practised
Southeast Asia	Species not identified	Damage to sweet potatoes is recorded in Sarawak and Laos
Oceania (South Pacific)	*Rattus rattus* ship rat; *Rattus exulans* Polynesian rat; *Rattus norvegicus* Norway rat	Damage occurs throughout the whole area. Generally very little control is practised against rodents in root crops
Australia and New Zealand	—	No information
West Indies	*Rattus rattus* ship rat; *Rattus norvegicus* Norway rat; *Dasyprocta aguti* agouti	Information on species is from the Cayman Islands where sweet potatoes, cassava and yams are grown

Region	Scientific and common names	Remarks
Latin America	*Dasyprocta variegata* agouti or siwayru	The native agouti or siwayru is said to 'demolish' root crops such as cassava and sweet potatoes.
	Dinomys braniku paca	The paca is a large river bank rodent which eats root crops but is much rarer than the agouti (Gade, 1975).
Atlantic Islands	*Rattus rattus* ship rat; *Rattus norvegicus* Norway rat	Sweet potatoes are attacked by these species on St. Helena.
Far East	—	Inadequate information.

*Unless otherwise indicated the information in this table is from Hopf *et al.,* 1976.

Where stores are sited in larger villages or towns it is more likely that the main pest will be one or more of the three cosmopolitan species; *Rattus rattus, Rattus norvegicus* and *Mus musculus.* These species are more flexible in their behaviour than most indigenous rodents and have become adapted to living in urban areas.

As well as damaging the stored crops, rodents damage the structure of stores, making holes in walls, roofs and floors. They may also cause serious economic loss by damaging the bags or packages in which foodstuffs are stored.

CONTROL IN THE FIELD

The particular method or combination of methods which is chosen to prevent rodent damage to root crops will depend on local economic, cultural and social factors. There are three main methods from which to choose. Of these, trapping has been employed since the earliest times, and much skill has been developed in its use. Farmers often understand less well the other two, poisoning and environmental alteration. It is these last two techniques, however, which have the biggest contribution to make to increasing crop yields.

There is a lack of good information about which rodent species are generally responsible for the damage. Thus the methods recommended should be looked upon as a guide to be modified as circumstances and experience dictate.

Poisons

There are three main kinds of poisons for rodent control, multi-dose, single-dose and gassing poisons. Warfarin, coumatetralyl, chlorophacinone, diphacinone, coumachlor, difenacoum and calciferol are multi-dose poisons. All except the last are anticoagulants, which have an effect on the blood system of the animal. It is these compounds which will normally give the best results — up to 100% kill — of rats in root crops. An example of a single-dose poison is zinc phosphide. Generally both multi-dose and single-dose poisons have to be mixed with a bait and have a lethal effect when eaten but gassing poisons release poisonous gases which are breathed in by the rodents.

In most situations where there is a rodent problem in growing crops, the use of a bait containing a multi-dose poison is the best method of control (Fig. 16). Whether a multi-dose or a single-dose poison is used, considerations as to the choice, the mixing and the application of bait are much the same and will be dealt with in the next section.

Fig. 16. Warfarin bait-block in plot of sweet potatoes in the Solomon Islands. (Pest Infestation Control Laboratory)

Baits

Rats and mice will usually eat cereals (either milled or whole grain) and often the cheapest locally available kind will be accepted as bait by the rats. Some grains, like whole maize, may be too hard and might be more acceptable as a meal. Rodents tend to like sugar, and a small amount added to a bait may make it more attractive. Sometimes a mixture of cereals will give a good bait.

Bait mixing

The poison must be mixed evenly throughout the cereal so that it is all poisoned. It is also very important that the correct amounts of poison and bait are mixed; follow the instructions on the label. The concentrations of rodenticides have been calculated to produce the maximum kill of animals. If too little poison is used, fewer animals will die and it may take longer to work; if too much is used, the poison bait may be unpalatable and too little will be eaten.

Many of the anticoagulants such as warfarin are sold as a master-mix or concentrate and require to be mixed in the proportion one part by weight of concentrate to 19 parts of bait. A useful amount that can be mixed in a bucket or bowl with a spoon, is 2 kg. When larger quantities are required they should be made up in 2 kg lots unless mechanical mixing equipment is available.

Safety: Always handle poisons carefully. Avoid breathing in poison dust and cover any cuts on the hands or arms. Buckets, spoons etc used for bait mixing should not be used for other purposes.

Mixing of zinc phosphide should be done immediately before use. Great care must be taken not to inhale the poisonous fumes of phosphine gas produced by zinc phosphide.

Suitable mixtures

If a cereal meal or flour is chosen as bait, a satisfactory mix can be obtained by mixing the poison and the meal together for about one minute.

Quantities for warfarin bait

Fine cereal meal or flour	19 parts (1900 g)
Warfarin concentrate (0.5%)	1 part (100 g)

Quantities for zinc phosphide bait

Fine cereal meal or flour	39 parts (975 g)
Zinc phosphide	1 part (25 g)

One kilogram is usually sufficient to mix at one time for small-scale treatments.

If a whole grain bait is chosen, it will be necessary to add a 'sticker' to ensure that the poison is evenly distributed. Corn oil or other vegetable oil is the best choice although a mineral oil is also acceptable.

Quantities for warfarin bait

Wheat or other grain	18 parts (1800 g)
Corn oil	1 part (100 g)
Warfarin concentrate (0.5%)	1 part (100 g)

Quantities for zinc phosphide bait

Wheat or other grain	36 parts (925 g)
Corn oil	2 parts (50 g)
Zinc phosphide	1 part (25 g)

Making wax blocks

A very useful way of making long lasting baits is by mixing the poison and the bait with melted paraffin wax. The wax sticks the bait and poison together and small blocks can be prepared by the following method:

Quantities for warfarin bait blocks

Fine cereal meal or whole grains	10 parts (1000 g)
Sugar	1 part (100 g)
Warfarin concentrate (0.5%)	1 part (100 g)
Semi-refined paraffin wax (m.p. 60–65°C)	8 parts (800 g)

The weights of ingredients suggested could probably be increased if wax block making is to be undertaken on any scale. If a metal bucket is to be used, 5 kg will probably be as much as can be well stirred, but up to 10 kg can be mixed at one time if a large enough vessel is available.

To prepare the bait blocks, first heat the paraffin wax just enough to melt it. It is important not to overheat the wax or the poison will dissolve in it and will not be digested by the rats. Next, stir in the cereal, sugar and poison until all are thoroughly mixed. Then pour the mixture into a flat tray and leave it to set. When set hard enough, cut into blocks each weighing about 100 g.

The advantage of wax blocks over loose baits lies in their greater resistance to the weather. They remain palatable to rats even when exposed to rain and sun.

Baiting and bait containers

Baits should be laid close to holes, runways or places where rodents are known to be living. The spacing of baits will depend on local circumstances, and may be from 1–15 m according to the number of rats. Unless poisoned wax blocks are used, the baits may need protecting against rain, by using containers. Bait containers may be wooden or metal boxes with entry holes cut out for the rats. They may also be hollow pipes, about 600 mm long. Pieces of plastic drain pipe would serve well; the bait should be placed mid-way along the inside of the pipe so it is safe from non-target animals and birds. If bamboo is available, pieces cut with the solid node at one end can also be used. To protect bait against heavy rain, bamboo containers should be placed sloping with the open end down. The essential need is for the bait to remain fresh and palatable to the rats, and as safely hidden from other animals as possible. When necessary, the bait containers should be marked with the words 'RAT POISON' or other appropriate words or symbols.

Using multiple-dose poisons

Anticoagulants are generally used at concentrations which make them more or less comparable in effectiveness. In most instances, therefore, the cheapest or most easily available should be selected. Many anticoagulants are marketed as ready-to-use baits that require no further mixing. Such bait is often expensive, and has the disadvantage of preventing variation in the bait constituents to suit different circumstances. Thus, it is generally best to use concentrates (master-mixes), which are marketed in various strengths, usually in the range 0.1–1.0% active ingredient. The concentrate can then be mixed with a locally available cereal bait to give the required final concentration. The mixing instructions on the label should be carefully followed to obtain the correct mix. Where, however, it seems unlikely that the user will heed such advice, it might be better to advocate a ready-mixed bait.

Anticoagulant baits act slowly over several days and it is not necessary to use unpoisoned bait first to get good results, as with single-dose poisons (see later). Baiting with anticoagulants should be generous; about 100–200 g at each place.

Once laid, the baits should be revisited every few days and more bait laid where necessary so that it is continuously available to the rats. Most rats will need three or four consecutive days feeding on the bait before they succumb, so it may easily be three or four weeks before all the rats have found the baits and been killed. A successful treatment is generally indicated when the bait ceases to be eaten.

Where crops are liable to be reinfested by more rats after the anticoagulant treatment, it may be useful to keep some of the baiting points permanently replenished with bait to deal with invaders. This problem of reinfestation after a good measure of control has been achieved can also be minimised if all the farmers in an area can be persuaded to carry out their rat control simultaneously.

Occasionally anticoagulant poison-baits fail to control rats. These failures are most often due to underbaiting, but sometimes an unattractive bait may be the cause, or positioning of bait points might have been so poor that the rats could not easily find them. It is also possible that too small an area has been treated and that reinvasion has taken place from an untreated area. In all these situations the remedy is self-evident, but if none of these possibilities seems to be the reason for failure, and the rats continue to eat the bait, it is reasonable to suspect resistance to the poison.

When resistance to warfarin, coumatetralyl, chlorophacinone, diphacinone or coumachlor is suspected, difenacoum should be used to poison the rats. Difenacoum, one of the newer anticoagulants, is very valuable in being effective against rats resistant to other anticoagulants.

If calciferol (vitamin D) is used to poison rats, follow the same procedure as for anticoagulant baiting. However, it is even more important that bait should be continuously available during the first few days of the treatment. Baiting should therefore be generous and the baits checked daily for the first three days. This is necessary because calciferol acts faster than anticoagulants on most rats. Any initial shortage of bait will tend to prevent some of them from eating a lethal dose, and there is a possibility that these surviving rats will then have an aversion to calciferol. Feeding on baits should normally stop after 7–10 days. If live rats are still present three weeks after the start of a calciferol treatment, use another poison on the survivors.

Using single-dose poisons

Of the several single-dose poisons that are generally available for rodent control, only zinc phosphide is safe enough for use by untrained personnel. It is cheap, widely available and many people already have some experience of using it. Its main disadvantage is the risk to other animals which might eat the poisioned bait or poisoned rats. Zinc phosphide bait will kill, on average, 40% of the rats on the first night of poisoning. Lay many small baits of 5–25 g. If it is safe to leave the baits down for further nights, more rats may be killed. Finally any bait that is not eaten should be removed and burnt or well buried.

Zinc phosphide acts rapidly and usually kills rodents after a single feed. However, if rodents start to feel ill before they have eaten a lethal amount, they may become poison shy and not feed again. For this reason, better kills (on average 80%) are obtained when unpoisoned bait ('pre-bait') is first put down for the rats, to condition them to feeding on it. If it is necessary to use bait containers, it may be an advantage to put them down some days before baiting starts. 100–200 g of unpoisoned bait should be laid at each place where it is thought the rats will eat, and where there are many rats, baits should be laid every few metres.

The bait should be checked every day for several days and more put down if necessary. When the rats are feeding freely, the unpoisoned bait should be removed and be replaced by the same bait, but now containing the poison, at all the points at which unpoisoned bait was eaten. To ensure that enough poison is laid, put down at each place, half as much bait as the amount taken on the last night of unpoisoned baiting, but not less than 25 g. Most rats will feed on the first night of poison baiting and the poison bait should usually be removed on the next day and burnt or well buried. If however, the bait can be safely left down longer further rats may be killed for up to a week.

Gassing poisons

Gassing powders act by releasing poisonous gases which are breathed in by the rodents. There are two main ones which have been used against rats; sodium cyanide and aluminium phosphide. They produce the poisonous gases hydrogen cyanide and phosphine respectively, when they come into contact with moisture.

Sodium cyanide gassing powder and aluminium phosphide tablets will not always give as good control as anticoagulant poisons, but they are often easily available, especially aluminium phosphide tablets which are used widely for insect control in stored grain.

The main use of gassing poisons is to control rodents living in burrows in well consolidated damp soil, around field edges for instance, where a good concentration of gas can be built up. A small spoonful of sodium cyanide gassing powder or two tablets of aluminium phosphide should be placed well inside each hole of the burrow system and the holes then blocked so that a lethal concentration of gas can build up. Holes re-opened by rodents should be gassed again the next day and the procedure repeated daily until all the holes remain blocked. Sodium cyanide powder may also be pumped into rat burrows, but this method is of less value as some rats bolt and so escape.

Safety: Gassing powders and tablets are extremely dangerous poisons if taken by mouth or inhaled. If accidents are to be avoided, great care is necessary in handling and storing them.

Trapping

Even when most skilfully used, traps will rarely control rats as effectively as poisons. Their use is probably best confined to larger rodents being caught for food or to the few survivors of a poison treatment.

Many different local traps will catch rats, and those such as the breakback trap (Fig. 17) which kill the animal, solve the problem of having to kill rats that are caught in live-

Fig. 17. Ship rat (*Rattus rattus*) caught in a breakback trap in a plot of sweet potatoes, Solomon Islands. (Pest Infestation Control Laboratory)

traps. Any locally available food that the rats will eat can be used for bait which should be placed in the centre of the treadle.

Weed and bush clearance

There is some evidence that weed control both within the crop and around the field edges will limit the numbers of rodents. This may be a very useful additional benefit to weed control done primarily for directly improving crop yields.

Bush clearance close to fields also limits rodents by reducing sources of food and living places, as many species do not actually live in the crops they attack, but in the undergrowth, banks and grassland surrounding it. Bush clearance at intervals may well be less costly than poisoning. It is not possible to give specific recommendations for how much bush must be cleared from around fields, but it is probable that a cleared band of several metres wide is needed.

CONTROL IN STORES

Proofing and hygiene

Making the store rodent-proof is probably the most satisfactory approach, but it is not always practicable and regular poisoning may be necessary to keep rodents in check. A very important and inexpensive method of minimising a rodent problem is scrupulous attention to hygiene in and around the store. All potential rodent living areas — piles of unwanted building materials, wooden packing cases, waste paper and floor sweepings — should be removed completely and their further accummulation prevented. In addition all undergrowth should be cut down and kept at a height which will enable any new rat holes and runs to be readily detected.

Poisons

The use of poisons has already been outlined in the section on control in the field. The same multi- or single-dose poisons can be used, but care must be taken that no contamination of the stored roots occurs. Aluminium phosphide tablets will kill rodents in stores, as will other fumigating gases but should be used only with very strict safety measures; gassing powders are too dangerous to use in or near buildings.

Trapping

Traps may be useful to get rid of the survivors of a poison treatment or to catch a few rodents getting into a store.

Bibliography

ACLAND, J. D. (1971). *East African crops.* pp. 252. Longman, London for FAO.

BJORNSON, B. F. *et al.* (1968). *Control of domestic rats and mice.* pp. 41. US Department of Health, Education and Welfare, Public Health Service, Rockville, Maryland, USA.

DAVIS, R. A. (1970). *Control of rats and mice. Bulletin* No. 181, pp. 28. HMSO, London.

GADE, D. W. (1975). *Plants, man and the land in the Vilcanota Valley of Peru.* pp. 240. Junk, The Hague.

HOPF, H. S., MORLEY, G. E. J. and HUMPHRIES, J. R. O. (Eds) (1976). *Rodent damage to growing crops and to farm and village storage in tropical and subtropical regions. Results of a postal survey 1972–73.* pp. 115. Centre for Overseas Pest Research and Tropical Products Institute, London.

HOWARD, W. E. and MARSH, R. E. (1974). Rodent control manual. *Pest Control* 42(8) Supplement. pp. 21.

JARVIS, J. U. M. and SALE, J. B. (1971). Burrowing and burrow patterns of East African mole-rats *Tachyoryctes, Heliophobius* and *Heterocephalus. Journal of Zoology, London* 163 (4): 451–479.

JENSON, A. G. (1965). *Proofing of buildings against rats and mice. Technical Bulletin* No. 12. pp. 18. HMSO, London.

PEST INFESTATION CONTROL LABORATORY (1976). *Control of rats and mice.* Reference manual for pest control personnel. Ministry of Agriculture, Fisheries and Food, Pest Infestation Control Laboratory, Tolworth, UK.

ROWE, F. P., WILLIAMS, J. M. and WILSON, E. J. (1975). *Rat control in the South Pacific. South Pacific Commission Handbook* No. 1. pp. 52. SPC Bureau, Sydney, Australia.

TAYLOR, D. K., DRUMMOND, D. C. and ROWE, F. P. (1970). Biology of rodents; Control of rodents. In *Food storage manual.* Prepared by Tropical Stored Products Centre, Slough. pp. 136–137 and 742–768. Food and Agriculture Organisation, Rome.

WALKER, E. P. (1975). *Mammals of the world.* 3rd Edition, Volume 2. pp. 647–1479. The Johns Hopkins University Press, Baltimore and London.

WILSON, E. J. (1972). *The assessment of rat damage. South Pacific Commission Information Document* No. 28. pp. 16. Noumea, New Caledonia.

STORAGE*

Introduction

Traditionally very little post-harvest storage of tropical root crops has been practised, and commonly the crops are simply left in the ground and harvested as and when needed and then immediately used or, as is sometimes the case with cassava, processed into a dried product with a longer natural storage life. Storage of a crop is required in those situations where the duration of the demand for it exceeds the period of production. In those areas where consumers only need to have produce available at times of production, for example, as a seasonal vegetable, there is little or no need for storage of the crop except in some cases for 'seed' purposes. This situation is not, however, to be confused with those in which consumers have been forced to accept only seasonal availability of a particular crop because of the absence of suitable storage systems. In those areas where continuous production of the crop is possible, due to favourable year round conditions, little or no storage is required. Thus it is seen that the need for and duration of storage required is largely determined by consumer demand and the production cycle of the crop. In any given area these factors should be well established and understood before detailed consideration is given to storage factors. Some simple traditional storage methods and also some improved and sophisticated systems have been developed in different areas of the world to meet consumer demand during non-production periods and some of these are briefly discussed below.

Any storage system, no matter how simple, costs money. Thus where storage is required to meet consumer demands, the consumers (or their supporters/sponsors) must be prepared to pay for the cost of the storage systems required to bridge the production gaps and satisfy the consumers requirements. The cost of storage not only involves the costs of constructing, maintaining and running the particular storage structures used, but must also take into account such factors as the value of the produce lost during storage and post-harvest handling operations in addition to consideration of delayed earnings and 'storage risk'.

Tropical root crops, with the exception of cassava, are generally stored in the fresh condition and some of the major factors influencing their storage life are discussed below. In addition to being 'stored' in the fresh state, cassava is frequently processed into a variety of dried products which have greater storage potential because they are essentially inert. Some of the factors which influence the storage of these dried products are discussed below.

---

*This chapter was written by Dr R. H. Booth, Tropical Products Institute, London.

STORAGE METHODS

In general the post-harvest storage of tropically produced root crops has not been practised on a large scale and thus the little information that is available usually relates to simple traditional methods. It is always important to realise that the most efficient method of storing and handling root crops is not absolute but is relative to a large number of continuously changing factors and conditions and therefore different methods will be more or less appropriate in different circumstances. In any particular situation the methods should be selected so as to yield maximum returns on the investment available.

Sweet potatoes

Generally in the tropics sweet potatoes are only held for short periods prior to marketing and a wide range of containers and storage buildings including caves are used. Also, structures similar to the clamps described below for cassava have been suggested for storage of sweet potatoes in those portions of the highlands of Papua New Guinea subject to periodic frosts.

Where cultivation of sweet potato has spread to semi-tropical and temperate zones, for example southern USA and Japan, the crop has to be harvested before it is damaged by frosts. Consequently, and as continuous production is not possible in these regions, improved and sophisticated methods have been developed for storing the tubers in order to extend their market life, to prevent gluts and shortages, and to preserve planting material for the following year. Thus it has been shown that ideally, after curing (see p. 49), sweet potatoes should be held at 13–15°C with a relative humidity of 85–90%. When stored in bulk a small amount of ventilation is needed to replace oxygen consumed and to remove carbon dioxide produced. However, in the majority of situations in the tropics, refrigeration will be required to obtain and maintain these ideal conditions. In many tropical regions the use of such sophisticated storage techniques as refrigeration will not be possible or even desirable in the near future for largely economic, managerial and sociological reasons, but where a long storage period is required to meet consumer needs such methods should be considered. Also, in those regions where prolonged storage is found necessary, the possibility of simple processing of the tubers into some form of acceptable dried food with an inherently longer storage life, as is practised with cassava roots in many regions, could be considered.

Cassava

Traditionally cassava is left in the ground and harvested when needed and then immediately used or processed into a dried product with a longer storage life. While the lack of any specific maturity period and thus the possibility of on-plant in-ground 'storage' is often advantageous in the case of cassava, this method of 'storage' can lead to problems in those regions where there is considerable pressure on land use. Also, susceptibility to pathogenic losses increases and palatability and extractable starch content decline when harvesting is very greatly delayed.

Small quantities of harvested cassava roots can be preserved for a few days using such simple traditional techniques as reburial in moist soil, coating in mud, and placing under water.

Recently it has been shown possible to considerably extend the storage life of quantities of cassava roots by placing them in clamps in the field or in containers packed with moist materials. While neither of these techniques has yet been widely applied they obviously merit further attention. Clamps are constructed by placing the roots in cone shaped heaps and covering them with straw and soil but at the same time providing a little ventilation (Figs. 18 and 19). The exact clamp design must be determined experimentally for each specific environment. Keeping roots packed in materials such as moist sawdust, moist coir dust, or moist peat in wooden, metal, or cardboard boxes has been shown to be very successful for periods of between two and six weeks and so should prove a valuable aid where cassava roots are marketed fresh (Fig. 20). Many of these simple cassava storage techniques provide conditions which promote curing and reduce moisture loss.

More sophisticated techniques such as refrigerated storage at temperatures below 6°C, and coating washed roots in paraffin wax have also been shown to be successful in maintaining roots in acceptable condition for periods of at least one month.

Fig. 18. Clamps constructed of earth and straw. (Booth (1977), *Experimental Agriculture* 13: 119–128)

Fig. 19. Cassava roots showing no deterioration after 8 weeks in a clamp compared with roots showing primary deterioration symptoms after 3 days storage at ambient conditions. (Booth (1977), *Experimental Agriculture* 13: 119–128)

Fig. 20. Cassava roots showing no deterioration after 8 weeks storage packed in boxes with moist sawdust. (Booth (1977), *Experimental Agriculture* 13: 119–128)

Where cassava roots are processed into a dried product immediately following harvesting it is important that all the operations in the process are well coordinated and carefully controlled and managed so as to avoid any hold ups or delays in processing. Cassava roots should be processed as soon as possible after harvesting, ideally within 24 hours; any further delays will result in a lowering of final product quality.

Yams

Yams, more than any other tropical root crops have regularly been stored by traditional methods in the tropics and in some cases these storage practices have developed as part of the culture of the local peoples. This is particularly true in the case of *Dioscorea rotundata,* the African white yam, in the yam producing areas of West Africa but also with *D. alata* in other regions such as the Trobriand Islands of Papua New Guinea.

During dry seasons yams can be left unharvested in the ground without appreciable loss of quality but are for several reasons, such as damage by pigs and rats, normally harvested for storage. Storage facilities vary among farmers throughout the yam growing areas. It is common to find yams stored in pits, in piles, or on the ground under houses but frequently slightly improved techniques are practised. They are frequently stored in heaps on the floor or on shelves or racks, the latter arrangement is preferable to ensure adequate ventilation and prevent the spread of rots. In the Ivory Coast small shelters are built in which the tubers are stacked to about one metre high; this method is said to favour scale insect infestation (see p. 172). Silos in which early crop yams are stored are built in the Ivory Coast and northern Ghana. Also, more elaborate yam barns are commonly used in West Africa. A yam barn is a shed with woven frames to which individual yams are tied. The frames are usually no more than 2 m high and several metres long. These structures provide adequate ventilation and protection from termites. In Oceania yams are sometimes stored standing vertically on a raised shelf beneath a thatched roof (Figs. 21 and 22).

Slightly improved storage methods involving first curing (see p. 49) followed by storage in a cool well-aired place have been developed to reduce storage losses. Refrigerated storage at temperatures below 10–20°C is not feasible as chilling damage occurs (see p. 43).

Different yam species and varieties differ very considerably in their storage characteristics and attention should be paid to these factors during the choice and selection of planting material. Also, greater attention could be given to the simple processing of certain yam species into dried products such as yam flakes and flour.

Taro and other aroids

Little or no scientific information is available on storage methods for aroids and most frequently they are used directly or marketed soon after harvesting. Reports on the storage life of aroids vary considerably but generally the cormels of *Xanthosoma sagittifolium* store better than corms or cormels of *Colocasia esculenta.* Corms of *Colocasia esculenta* may, however, be stored for up to four weeks in soil pits lined with leaves provided that the pits are situated in shaded, well-drained soil. It has been suggested that the stimulation of root and shoot growth which occurs subsequent to

Fig. 21. West African yam barn. (E. Terry)

Fig. 22. Pacific Island yam barn. (COPR)

burial in soil also enhances healing of wounds. Using this method of storage there is little deterioration of corm taste and texture as has been found to occur during simple shed storage, but bacterial soft rots are frequently a problem and will probably limit long-term storage by this method. Also corms and cormels of *Colocasia* and *Xanthosoma* can be successfully kept in good eating condition for up to four weeks when simply placed in polythene bags. In both soil pits and polythene bags it is believed that curing (see p. 49) is promoted and excessive moisture loss prevented.

It is considered that many of the simple traditional methods described above and employed for storing the various root crops could be improved by paying greater attention to the quality, particularly physical condition of the produce placed into storage and by subjecting the produce to a wound healing or curing process (see p. 49). Also, for short-term storage and as an aid to marketing it is frequently possible to keep root crops in good condition in closed polythene bags or stored packed in moist materials such as peat, sawdust or coir dust. For long-term storage care must be taken to avoid condensation of free moisture on the produce as this will stimulate rapid microbial rotting. Wherever possible, it is advisable to keep simple records of the air temperature amongst the stored produce; a sudden and large rise in the temperature is an indication of rotting or other storage problems and it is advisable to utilise or market the produce immediately. Greater consideration should also be given to simple processing of root crops and the production of dried products with greater storage potential.

STORAGE OF FRESH PRODUCE

Storage losses: nature and magnitude

During the storage and handling of fresh root crops losses are due to physical, physiological, or pathological causes or various combinations of all three.

Physical damage

Only too frequently all root crops are treated and handled as inert objects instead of living organs and very considerable losses result directly from physical injury and indirectly through enhanced physiological and pathological losses. Damaged produce will always have a shorter post-harvest life than undamaged produce.

In addition to being caused by rough handling physical damage can also be the result of exposure to both high and low extremes of temperature. Most tropically produced root crops, but notably yams and sweet potatoes, are subject to chilling damage at temperatures below 10–12°C. Chilling injury is expressed in many ways but commonly involves internal discolouration and tissue breakdown, increased susceptibility to decay and impaired culinary quality. Root crops exposed to very high temperatures, such as those caused by direct sunlight, suffer a similar type of damage.

Physiological damage

As root crops are living organs metabolic losses such as wilting and respiratory loss of dry matter always occur. The rate and magnitude of such losses will be much influenced by the storage environment, particularly temperature and relative humidity, but as indicated above damaged produce will lose water more rapidly, and frequently have a higher respiration rate than sound produce.

Additional physiological losses, in all cases except cassava, can occur as a result of sprouting which in general renders the produce less marketable and also accelerates water and respiratory losses. Sprouting usually occurs following a natural dormant period which may vary considerably between varieties and which is also influenced by storage environment and produce condition. Damaged produce frequently sprouts earlier than sound produce.

In cassava considerable losses arise from vascular streaking/primary deterioration which normally occurs within three to five days of harvesting and which renders the roots unmarketable. The exact cause of this rapid post-harvest deterioration has not yet been determined but as it is not directly associated with attack by any single pathogen it may be described as physiological in nature. This rapid internal discolouration usually starts at the sites of damage and first appears in the vascular tissues but later spreads throughout the root.

Physiological changes in quality of all root crops may occur in storage, frequently due to changes in the starch–sugar equilibrium which can be influenced by storage temperature. In general, storage at low temperatures increases the sugar content, resulting in sweetening. Sprouting is also usually accompanied by an increase in sugar content. In some circumstances this sweetening is regarded as a desirable characteristic as in sweet potatoes but in other root crops it is generally considered detrimental.

Pathological damage

Attack by microorganisms is probably the most serious cause of post-harvest loss in root crops although it must be remembered that both physical and physiological damage frequently predisposes material to pathogen attack. In particular, many truly post-harvest fungal pathogens are wound parasites and only capable of attacking storage organs at sites of injury. Quantitative pathogenic losses result from the breakdown of host tissues by microorganisms. The pattern of attack is frequently an initial infection, either pre-harvest or commonly through a post-harvest wound, by one of a few specific pathogens, followed by a massive attack by a broad spectrum of organisms which grow on the dead or moribund tissues. Qualitative pathogenic losses are typically the result of blemish diseases which render the produce less attractive and so reduce its market value.

Examples of pathogens which are frequently encountered causing post-harvest losses are listed in Table 5.

TABLE 5. COMMON POST-HARVEST ROOT CROP DISEASES

Crop	Disease	Pathogen	*CMI Descriptions of Pathogenic Fungi and Bacteria*
Sweet potato	Black rot	*Ceratocystis fimbriata*	141
	Blue mould rot	*Penicillium* sp.	
	Charcoal rot	*Macrophomina phaseolina*	275
	Dry rot	*Diaporthe batatatis*	
	Foot rot	*Plenodomus destruens*	
	Grey mould rot	*Botrytis* sp.	
	Java black rot	*Diplodia tubericola*	
		Botryodiplodia theobromae	519
	Mucor rot	*Mucor racemosus*	107, 529
	Ring rot	*Pythium* spp.	
	Scurf	*Monilochaetes infuscans*	
	Soft rot	*Rhizopus* spp.	
	Soil rot	*Streptomyces ipomoeae*	
	Surface/end rot	*Fusarium* spp.	
	Internal cork	Virus ?	
Cassava	White root disease	*Rigidoporus lignosus*	198
	Phytophthora root rot	*Phytophthora* spp.	
	Rosellinia root rot	*Rosellinia necatrix*	352
	Sclerotium root rot	*Corticium rolfsii*	410

Crop	Disease	Pathogen	*CMI Descriptions of Pathogenic Fungi and Bacteria*
	Cassava bacterial blight	*Xanthomonas manihotis*	559
	Post-harvest vascular streaking	'primary deterioration'—physiological	
	General post-harvest decay	'secondary deterioration' species of: *Pythium, Fusarium, Rhizopus, Rhizoctonia, Aspergillus, Penicillium, Botryodiplodia, Diplodia, Bacillus, Erwinia, Corynebacterium* and others	
Yams	Dry rot	*Corynebacterium* sp.	
	Watery rot	*Erwinia* sp.	
	Soft rot	*Botryodiplodia theobromae*	519
		Fusarium spp.	
		Penicillium spp.	
		Penicillium sclerotigenium	
		Macrophomina phaseolina	275
		Aspergillus niger	94
		Rhizopus sp.	
		Trichoderma viride	
		Hendersonula toruloidea	274
		Rosellinia bunodes	351
		Botrytis sp.	
		Colletotrichum sp.	
		Serratia sp.	
		Rhizoctonia solani	
	Internal brown spot	Virus ?	
Aroids	Corm rot	*Botryodiplodia theobromae* (Figs. 23 and 24)	519
		Botrytis spp.	
		Fusarium spp.	
		Pythium spp.	
		Rhizoctonia sp.	
		Erwinia sp.	
	Java black rot	*Diplodia* sp.	
	Powdery grey rot	*Fusarium solani* (Fig. 25)	29
	Sclerotium rot	*Corticium rolfsii* (Fig. 26)	410
	Soft rot	*Erwinia* sp.	
		Phytophthora colocasiae	
		Pythium splendens	120

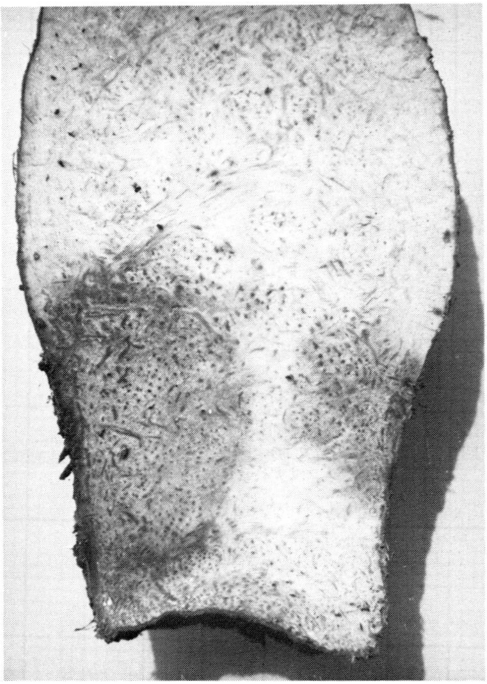

Fig. 23. Early stage of spongy black rot (*Botryodiplodia theobromae*) after seven days in storage. (Gollifer and Booth (1973), *Annals of Applied Biology* 73(3): 349–356)

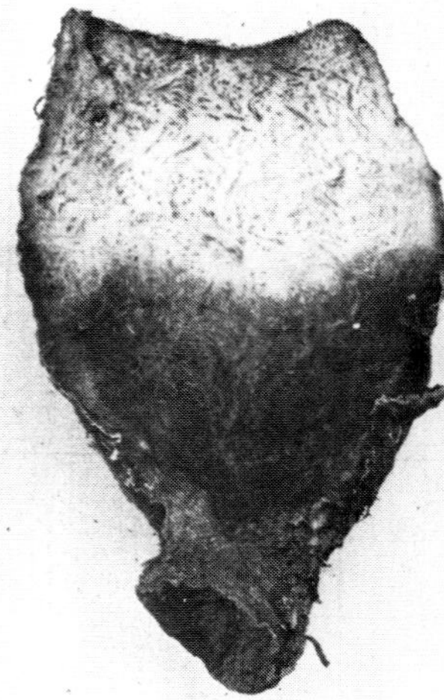

Fig. 24. Late stage of spongy black rot (*B. theobromae*) after 21 days in storage. (Gollifer and Booth (1973), *Annals of Applied Biology* 73(3): 349–356)

Insect pests

During the storage and handling of fresh root crops insect pests are not commonly the cause of any serious loss. The major exception to this statement is the sweet potato weevil, *Cylas formicarius,* which wherever present causes very serious damage to both the growing crop and stored tubers. Some storage damage to yams and aroids may also be caused by beetles, for example *Heteroligus* spp. and *Papuana* spp. Mealybugs can cause serious damage to yams (see p. 175) in storage, as a result of which the planting pieces do not germinate. The nematode *Pratylenchus coffeae* has also been shown to cause some damage in stored yams.

In addition to insect pests, considerable loss may be caused by rodents, particularly rats, and other animal pests such as monkeys and pigs.

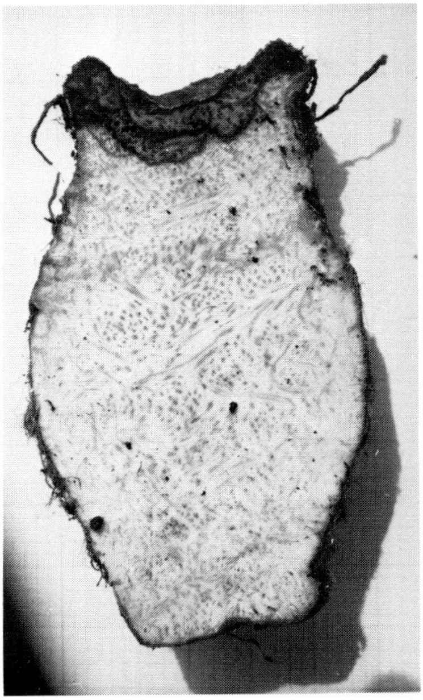

Fig. 25. *Fusarium* dry rot *F. solani* of taro corm after seven days in storage. (Gollifer and Booth (1973), *Annals of Applied Biology* 73(3): 349–356)

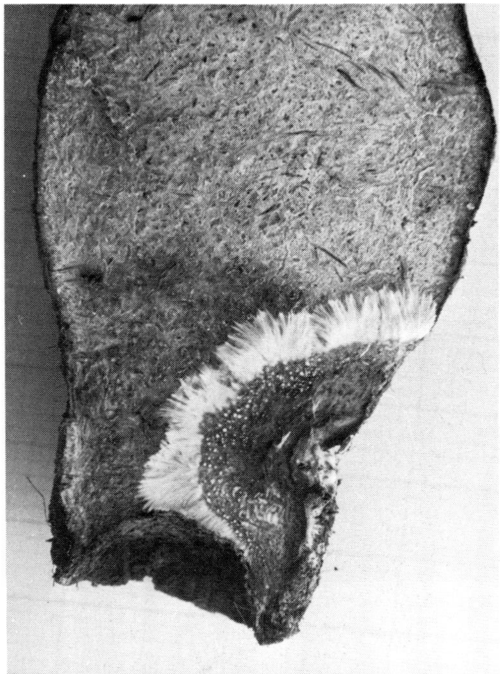

Fig. 26. *Sclerotium* rot (*Corticium rolfsii*) of taro corm after seven days in storage and one day's exposure of the cut surface. (Gollifer and Booth (1973), *Annals of Applied Biology* 73(3): 349–356)

The magnitude of post-harvest losses caused by the factors discussed above varies considerably, but on a global basis has been conservatively estimated at 25% of production. Using this as an average figure, the very large quantities of produce that are lost globally are shown in Table 6.

Prevention and reduction of storage losses

Post-harvest losses may be reduced by various physical and chemical means, some examples of which are given below, and also by using the produce immediately after harvest and so avoiding the need for storage.

In addition to specific post-harvest control measures, post-harvest losses may also be influenced by certain pre-harvest factors. For example, varieties may differ in several storage characteristics such as length of dormancy and disease resistance, cultural practices such as plant population and harvesting time and method can considerably influence the condition of the produce at harvest time, and certain pests and diseases which attack the storage organs before harvest are best controlled in the field.

TABLE 6. GLOBAL PRODUCTION AND POST-HARVEST LOSSES OF TROPICAL ROOT CROPS (MILLION TONNES)

Crop	Global production*	Post-harvest losses†
Sweet potatoes	134	34
Cassava	105	26
Yams	19	5
Aroids	4	1
Total	262	66

*FAO, 1974 †Coursey and Booth, 1972

Loss avoidance

In some tropical regions the problems of root crop storage can to a large extent be avoided as, due to favourable temperature and rainfall distribution, the crops can be produced continuously throughout the year.

In the specific case of cassava, which has no clearly defined maturity period, the crop can simply be left growing and only harvested as and when required. In general, this is a very satisfactory situation although palatability and extractable starch are reported to decline and susceptibility to root rots to increase with prolonged in-ground 'storage'. Also, in areas where the pressure on land is great this storage system is not satisfactory.

Physical control measures

The use of reduced temperatures, frequently involving refrigeration, is probably the most universally successful means of reducing post-harvest losses of most perishable agricultural or horticultural produce. However, the use of refrigeration for the storage of relatively low value commodities such as root crops is at present limited, particularly in the tropics. Also, as mentioned above, the scope of low temperature storage is limited in the case of yams and sweet potatoes by the dangers of chilling injury. Systems other than the use of refrigeration, where slightly reduced storage temperatures are obtained, for example those making use of cooler night air for ventilation purposes, are in some circumstances adequate for short-term storage. In some areas, caves in which temperatures may be well below ambient are used for the storage of cured sweet potatoes.

Reduced temperature storage is effective as metabolic processes slow down with a decline in temperature and so losses caused by respiration and sprouting can be very much reduced. Similarly, the metabolism of pathogens is slowed down and so rotting is frequently arrested during low temperature storage.

In certain cases, high temperature treatments of produce can be used to control storage diseases and to disinfect stores and storage containers. In particular, black rot and scurf of sweet potatoes may be controlled by exposing the roots for six hours at 50°C.

A considerable reduction in losses can be achieved by improving all handling operations and techniques in order to reduce mechanical injury. In addition to reducing the occurrence of damage, the effects of injury may be minimised by a simple wound healing or curing process during which general skin strengthening also occurs. Curing involves, first, suberisation of cell walls adjacent to wounds and is followed by the development of a wound periderm around the sites of injury and which is effective in retarding water loss and providing a physical barrier against infection by a large number of post-harvest wound pathogens. It is now well established with the majority of root crops that curing is one of the simplest and most effective ways of reducing post-harvest losses. The conditions of relatively high temperatures and humidities required to promote curing are given in Table 7. These conditions are generally readily obtainable by simply covering the produce and so slightly reducing the natural ventilation so that the heat produced by respiration and moisture evaporation from wounds and adhering soil is allowed to build up. Care must, however, be taken to avoid condensation of free water on the produce as this will encourage rapid bacterial decay. It is better to cover produce with materials such as straw and dried grass or at least to place such materials between the produce and tarpaulin or polythene type covers. Also, as oxygen is required in the curing process, a complete inhibition of ventilation, which would lead to carbon dioxide accumulation and oxygen depletion, should be avoided.

Curing must, if it is to be successful, be undertaken immediately after each handling operation, during which extensive damage occurs, otherwise pathogens will become established in the wounds before the periderm or barrier is formed.

TABLE 7. CONDITIONS REQUIRED FOR CURING ROOT CROPS

Crop	Temperature (°C)	Relative humidity (%)	Time* (days)
Cassava	30–35	80–95	4–7
Sweet potatoes	30–32	85–90	4–7
Yams	32–40	90–95	1–4
Aroids	Exact conditions not yet determined but probably similar to those above		

*The exact length of time required for proper curing varies with many factors such as variety, type of wound, maturity etc., but on average 7–14 days is recommended in practice.

Chemical control methods

Before fungicides and bactericides can be successfully used to reduce post-harvest disease losses, a thorough knowledge of the etiology and epidemiology of the diseases involved is required. Certain pathogens which cause post-harvest loss do, for instance, attack the storage organs while they are still attached to the growing plant prior to harvest and so chemical and other control measures are best directed at reducing the incidence of these diseases in the field. Examples of such diseases include: black rot of sweet potatoes (see p. 61); white root disease (see p. 110), *Phytophthora* (see p. 109), *Rosellinia* (see p. 112), and *Sclerotium* root rots (see p. 112) and cassava bacterial blight (see p. 113) of cassava; dry rot (see p. 157) and watery rot (see p. 155), of yams and soft rot of aroids (see p. 184). Other pathogens only attack the storage organs at or after harvest, frequently through sites of injury as discussed above. The success of post-harvest chemicals in controlling the first described group of diseases is obviously extremely limited as the pathogens are already well established in the tissues at the time of harvest. Also the successful post-harvest use of chemicals for the control of the second group of diseases is limited although chemicals such as sodium-orthophenyl-phenate, dichloronitroaniline, borax, captan, thiabendazole and benomyl have all, in specific cases, been successfully used to control a broad spectrum of truly post-harvest pathogens and secondary biodeteriogens.

The post-harvest application of insecticides is rarely practised during root crop storage although in weevil infested areas roots of sweet potatoes may be dusted with DDT which must subsequently be washed off during preparation for market.

Rodents and other animals can be a serious storage problem and damage by them may in general be prevented by one of two approaches; either regular disinfestation in and around the store or through animal proofing of the store to prevent the entry of these pests. Proofing is probably the more satisfactory approach but it is not always practicable and regular poisoning may be necessary to keep animal pests in check. Chronic or acute poisons can be used in stores but great care must be taken to prevent contamination of the food products.

In addition to chemical disinfestation of stores against animal attack, fungicides and insecticides such as formalin and DDT may be used to fumigate stores and storage containers. This, however, is frequently not necessary provided that the store, storage containers and all handling equipment are kept clean.

A further group of chemicals which are occasionally used to reduce storage losses are sprout suppressants. While the use of such chemicals as α-naphthaleneacetic acid, tetrachloronitrobenzene, isopropyl N-phenylcarbamate, isopropyl N-chlorophenyl-carbamate, 3,5,5,-trimethylhexanol and maleic hydrazide is common in the storage of white potatoes their effectiveness in reducing sprouting losses of tropical root crops under the generally less sophisticated storage conditions is poorly understood and merits further investigation.

A further important consideration and limitation in the use of any post-harvest chemical application is that of residue levels and food additive legislation or control. Before being added to food crops all chemicals should have been rigorously screened and only used in strict accordance with the manufacturers recommendations and the food additive regulations of the country or countries concerned.

General handling and storage recommendations

1. Adequate attention should be paid to pre-harvest factors such as varietal selection and cultural practices which influence the post-harvest life of the produce.
2. Produce should be handled with extreme care so as to reduce physical injury to a minimum.
3. Avoid exposure of produce to extremes of temperature, for example direct strong sunlight.
4. The storage 'house' and storage containers must be clean.
5. Roots showing disease symptoms and severe mechanical injury should be discarded. The condition and health of the produce at the time it is placed in store is usually one of the most important factors contributing to the success or failure of storage.
6. Wherever possible subject produce to a wound curing period prior to longer-term storage.
7. Regularly inspect stored produce and use or market at the first sign of trouble.
8. Wherever possible keep records of the air temperature amongst the stored produce. A rise in temperature and over-heating of the produce is a good indication of trouble.
9. Only use post-harvest chemicals with extreme caution and due respect to food-additive and residue level control and legislation.
10. Remember that the most efficient method of storing and handling root crops is not absolute but is relative to a large number of continuously changing factors and conditions and therefore different methods will be more or less appropriate in different circumstances. In any particular situation select the methods which yield maximum returns on the investment available.

STORAGE OF DRIED PRODUCE

As cassava is the only major tropical root crop discussed here which is regularly processed into dried products comments will be confined to this crop although the principles underlying these comments could equally be applied to the storage of dried products produced from any other root crop. Many dried products, including chips and pellets, flour, starch, tapioca pearl and flake, and gari and farinha are produced from cassava roots in a variety of ways in different regions largely according to local or export needs, tastes and traditions. The different end uses and means of production of these products will not be discussed here. However, it must be remembered that due to the rapid post-harvest deterioration of cassava roots, harvested roots should be processed as soon as possible and that any delays will result in a lowering of the final product quality, often shown as a loss of whiteness which will reduce market value.

The problems encountered in the storage of the various dried cassava products are generally very different from those found during fresh produce storage and are more closely related to those of grain crops storage. As essentially inert products are involved, storage losses largely result from the activities of external biodeteriogens. Generally, attack by insects is the most serious storage problem, although in the many situations where the product is improperly dried, moulding can cause serious problems.

Different types of cassava product appear to attract different insect pests, although several of the more ubiquitous stored product pests are found on all types. A summary of insect pests found on cassava products is given in Table 8.

TABLE 8. INSECT PESTS INFESTING CASSAVA PRODUCTS IN STORAGE*

Species	Common name	Chips	Flour	Tapioca
Ahasverus advena	Foreign grain beetle	++	+	+
Alphitobius spp.	Lesser meal worm beetles, black fungus beetles			+
Araecerus fasciculatus	Coffee bean weevil, cacao weevil	++		+
Bostrychoplites cornutus		+		
Carpophilus spp.	Dried fruit beetles			+
Cathartus quadricollis	Squarenecked grain beetle	+		+
Coninomus constrictus				+
Cryptolestes spp.	Rust red grain beetles, flat grain beetles	+	+	
Dermestes spp.	Hide beetles			+
Dinoderus spp.		+	+	
Ephestia cautella	Tropical warehouse moth	+	+	+
Gnathocerus cornutus	Broadhorned flour beetle	+	+	
Heterobostrychus brunneus		+		
Lasioderma serricorne	Cigarette beetle	+	+	+
Lepidoptera species	Moths			+
Necrobia rufipes	Red-legged ham beetle, copra beetle		+	+
Oryzaephilus mercator	Merchant grain beetle		+	+
Oryzaephilus surinamensis	Saw-toothed grain beetle			+
Palorus ratzeburgi	Small-eyed flour beetle		+	+
Plodia interpunctella	Indian meal moth	+		
Pyralis manihotalis		+		
Rhizopertha dominica	Lesser grain borer	++	+	+
Sitophilus oryzae	Lesser rice weevil	++	+	+
Stegobium paniceum	Biscuit beetle, drug store beetle	++		+
Tenebroides mauritanicus	Cadelle	+	+	+
Thaneroclerus buquetii			+	
Tribolium castaneum	Rust red flour beetle	++	++	++

*Ingram and Humphries, 1972.
++Indicates major pests, + Indicates pests of infrequent occurrence.

A reduction in insect pest losses can be achieved in a number of ways, by far the simplest and most practical being the use of sealed containers which prevent insect infestation. In all cases only clean and if necessary disinfested containers and storage buildings should be used. Chemical means that have been recommended to reduce loss in infested stores include the use of gamma-BHC smoke bombs and fogging with DDT. Also the use of gamma-BHC/dieldrin impregnated jute bags has been shown to lower infestation levels. However, the use of DDT and dieldrin as suggested above could present health hazards and could not generally be recommended. Fumigation of bagged, infested chips with methyl bromide, ethylene dibromide and a mixture of ethylene dibromide and carbon tetrachloride has been shown to give rapid and effective control and residue levels were found to be so low as to be hazardless.

The literature reports that cassava chips prepared from the more bitter cassava varieties store longer before becoming infested and that gari, a dried fermented product, rarely suffers insect infestation and may in fact possess repellent qualities. The literature records no problems of insect infestation in cassava starch. Such factors warrant further investigation and scientific study.

All cassava products if not adequately dried are very prone to attack by various moulds. Similarly, in humid environments cassava products, if not stored in sealed containers, rapidly absorb moisture from the atmosphere and soon reach equilibrium moisture content. At a temperature of 30°C the equilibrium moisture content of coarsely ground cassava chips has been shown to be approximately 11, 12, 14, 16, and 18% at relative humidities of 45, 54, 64, 75 and 84% respectively. Cassava products are usually regarded as 'safe' and fit for storage at moisture contents of 12—13% and below. Thus in the majority of tropical environments it is necessary to store dried cassava products in moisture proof containers such as polythene bags or bags with polythene liners. Common moulds which have been identified include species of *Aspergillus, Mucor, Penicillium,* and *Rhizopus.* The appearance of moulding is always accompanied by loss in quality and palatability of the product and the danger of mycotoxin accumulation also exists. Provided the relative humidity is below 15%, carboxylic acids may be applied at 0.3% (3 *l*/t) to cassava chips to protect them from fungi and yeasts.

When adequate precautions to avoid insect infestation and moisture reabsorption are taken, well prepared dried cassava products have good keeping qualities and can frequently be simply stored for periods of twelve months or more.

Bibliography

ADENIJI, M. O. (1970). Fungi associated with storage decay of yam in Nigeria. *Phytopathology* 60: 590–592.

ANON (1973). Preservacion de Yuca fresca por el metodo de parafinado. *Teconolgia,* Instituto de Investigaciones Teconologica, Bogota, Colombia 86: 1–15.

BOOTH, R. H. (1974). Post-harvest deterioration of tropical root crops: Losses and their control. *Tropical Science* 16: 49–63.

BOOTH, R. H. (1975). *Cassava storage.* Series EE 16, Centro Internacional de Agricultura Tropical, Cali, Colombia. pp. 18.

BOOTH, R. H. (1976). Storage of fresh cassava (*Manihot esculenta*) I. Post-harvest deterioration and its control. *Experimental Agriculture* 12: 103–111.

BOOTH, R. H. (1977). Storage of fresh cassava (*Manihot esculenta*) II. Simple storage techniques. *Experimental Agriculture* 13(2): 119–128.

BOOTH, R. H., DE BUCKLE, T. S., CARDENAS, O. S., GOMEZ, G. and HERVAS, E. (1976). Changes in quality of cassava roots during storage. *Journal of Food Technology* II: 245–264.

COURSEY, D.G. (1967). *Yams.* pp. 240. Longmans, London.

COURSEY, D. G. and BOOTH, R. H. (1972). The post-harvest phytopathology of perishable tropical produce. *Review of Plant Pathology* 51: 751–765.

ECKERT, J. W. and SOMMER, N. F. (1967). Control of diseases of fruits and vegetables by post-harvest treatment. *Annual Review of Phytopathology* 5: 391–432.

EKUNDAYO, J. A. and DANIEL, T. M. (1973). Cassava rot and its control. *Transactions of the British Mycological Society* 61: 27–32.

GOLLIFER, D. E. and BOOTH, R. H. (1973). Storage losses of taro corms in the British Solomon Islands Protectorate. *Annals of Applied Biology* 73: 349–356.

GONZALEZ, M. A. and DE RIVERA, C. (1972). Storage of fresh yam (*Dioscorea alata*) under controlled conditions. *Journal of Agriculture, University of Puerto Rico* 56: 46–56.

HARTER, L. L. (1916). Storage rots of economic aroids. *Journal of Agricultural Research* 6: 549–574.

INGRAM, J. S. and HUMPHRIES, J. R. O. (1972). Cassava storage—a review. *Tropical Science* 14: 131–148.

JACKSON, G. V. H. and GOLLIFER, D. E. (1975). Storage rots of taro (*Colocasia esculenta*) in the British Solomon Islands. *Annals of Applied Biology* 80: 217–230.

KELENY, G. P. (1965). Sweet potato storage. *Papua and New Guinea Agricultural Journal* 17: 102–104.

KUSHMAN, L. J. (1971). Effect of injury and relative humidity during curing on weight and volume loss of sweet potatoes during curing and storage. *HortScience* 10: 275–277.

KUSHMAN, L. J. and WRIGHT, F. S. (1969). *Sweet potato storage. Agriculture Handbook* No. 358, USDA. 35 pp.

LUTZ, J. M. and SIMONS, J. W. (1958). *Storage of sweet potatoes. Farmers Bulletin* No. 1442, USDA. 42 pp.

MARTIN, F. W. (1976/7). *Tropical yams and their potential. Agriculture Handbooks* Nos. 457, 466, 495 and 502. USDA, Washington, USA.

MARTIN, W. J. (1971). Evaluation of fungicides against sweet potato black rot fungus, *Ceratocystis fimbriata. Plant Disease Reporter* 55: 523–526.

MINGES, P. A. and MORRIS, L. L. (1953). *Sweet potato production and handling in California. California Agricultural Experimental Station, Circular No.* 431. pp. 38.

MORRIS, L. L. and MANN, L. K. (1955). Wound healing, keeping quality and compositional changes during curing and storage of sweet potatoes. *Hilgardia* 24: 143–183.

OGUNDANA, S. K. (1972). Investigations into the causes of yam tuber rot in storage in Nigeria. *Journal of the West African Science Association* 17(2): 113–120.

OGUNDANA, S. K., NAQVI, S. H. Z. and EKUNDAYO, J. A. (1970). Fungi associated with soft rot of yams (*Dioscorea* spp.) in storage in Nigeria. *Transactions of the British Mycological Society* 54(3): 445–451.

OGUNDANA, S. K., NAQVI, S. H. Z. and EKUNDAYO, J. A. (1971). Studies on soft rot of yams in storage. *Transactions of the British Mycological Society* 56(1): 73–80.

OKAFOR, N. (1966). Microbial rotting of stored yams (*Dioscorea* spp.) in Nigeria. *Experimental Agriculture* 2: 179–182.

PASSAM, H. C., READ, S. J. and RICKARD, J. E. (1976). Wound repair in yam tubers: the dependence of storage procedures on the nature of the wound and its repair. *Tropical Science* 18: 1–11.

PASSAM, H. C., READ, S. J. and RICKARD, J. E. (1976). Wound repair in yam tubers: physiological processes during repair. *New Phytologist* 77: 325–331.

RIVERA, J. R., GONZALEZ, M. A., COLLAZO DE RIVERA, A. and CUEVAS-RUIZ, J. (1974). An improved method of storing yam (*Dioscorea alata*). *Journal of the Agricultural University of Puerto Rico* 58: 456–465.

THOMPSON, A. K., BEEN, B. O. and PERKINS, C. (1973). Nematodes in stored yams. *Experimental Agriculture* 9: 281–286.

WEIMER, J. L. and HARTER, L. L. (1921). Wound-cork formation in sweet potato. *Journal of Agricultural Research* 21: 637–647.

SWEET POTATO

DISEASES

Introduction

Sweet potatoes are affected by several different diseases in the field, some of which also affect the storage life of the tubers. Virus diseases are probably responsible for the greatest field losses and some fungus diseases can also cause economic losses. The use of clean planting material is the most important disease control measure and crop rotation aids control of many fungus diseases. The distributions given for each disease may have some omissions owing to the lack of written reports on the sweet potato disease situation in some areas.

FUNGAL DISEASES

Phaeoisariopsis bataticola, Pseudocercospora timorensis, Cercospora ipomoeae

Cercospora leaf spots

Distribution

Phaeoisariopsis bataticola: India, USA and Venezuela.

Pseudocercospora timorensis: Brunei, Fiji, Formosa, Ghana, Guinea, Hong Kong, India, Malaysia, Mauritius, Nepal, New Guinea, New Hebrides, Sabah, St Lucia, Sierra Leone, Solomon Islands, Sudan, Tanzania and Uganda.

Cercospora ipomoeae: Very common throughout the tropics.

Symptoms

Symptoms produced by these pathogens are generally similar. Spots are small (about 6 mm in diameter) and angular with brownish colouration. Leaves may drop from the lower parts of the stem in severe attacks. Damage to leaves is most severe during wet weather conditions.

Development and spread

Conidia are produced abundantly on the leaf spots and are spread by wind and rain. Generally no significant yield reductions are recorded.

Control

Control of these spots is not usually necessary; however, a trial in India showed that spraying every 15 days with zineb gave effective control and significant yield increases.

Elsinoe batatas

Scab

Distribution

Brazil, Indonesia, Japan, Malaysia, Pacific Islands (CMI Distribution Map 447).

Symptoms

Scabby lesions or small spots appear on the main veins and midrib on the undersurface of the leaves, which may be severely distorted and reduced in size. Lesions, which are grey to dark brown also occur on petioles and stems, causing distortion and twisting. The growing points of the vines may be killed.

Development and spread

Scab is common throughout the South Pacific, especially in the cool season. Spores are produced on the surface of the scabby areas and are dispersed by rain to infect healthy shoots. Temperatures between 13 and 26°C in the cool season are most conducive to disease development.

Control

Resistant varieties, if available, should be used in areas where scab is likely to be a problem. Various fungicides have been suggested for control. Benomyl at 1.04 g a.i./*l* water or copper oxychloride at 2.08 g a.i./*l* are recommended as is maneb at 2.08 g a.i./*l* every 10–14 days.

Fusarium oxysporum f. sp. *batatas*

Stem rot

Distribution

This disease appears to occur in temperate regions and is found in the USA, Japan and northern India.

Symptoms

Fusarium oxysporum f. sp. *batatas* (*CMI Descriptions of Pathogenic Fungi and Bacteria* No. 212) causes a vascular wilt (Figs. 27 and 28). The first symptom of the disease is a yellowing of the youngest leaves. The vascular system becomes blackened causing stems to darken internally. This darkening may extend several feet along stems. Infected plants wilt, collapse and die. Infected tubers have a blackened ring about 6 mm under the skin and develop a surface rot with shallow, sunken, circular lesions in storage. Plants produced from these tubers are likely to be diseased.

Stem rot is widespread in the USA and 10–50% losses in susceptible varieties may occur. Losses of up to 95% have been recorded.

Development and spread

The main sources of inoculum are infected tubers and plants. The fungus can live for several years on plant remains in the soil and can be spread locally by water, wind, animals, machinery etc. Diseased plants develop as a result of either diseased planting material or healthy plants becoming infected from the soil.

The incidence of the disease varies with the soil type, which may influence disease severity more than the inoculum density.

Conidia, which are readily dispersed, are produced on the dead leaves and stems.

Control

The most important means of controlling stem rot culturally are to use healthy planting material and crop rotation. Crop rotation will not eradicate the fungus but it will help to reduce the losses. Tobacco should be avoided in the rotation as some varieties are attacked by the stem rot fungus. Sweet potato should not be grown in the same ground more often than once in 3–4 years.

Stem rot resistant varieties of sweet potato are available in some areas. Local extension officers should be consulted for information on the availability of suitable varieties for a particular area.

Surface disinfection of infected roots will not control stem rot neither can the disease be controlled in infected vines by fungicidal dipping at the time of field planting. However the systemic fungicide thiabendazole has shown promise as a dip under experimental conditions.

Fig. 27. Resistant (right), intermediate (centre) and susceptible (right) sweet potato plants to Fusarium wilt (USDA, US Vegetable Laboratory, Charleston, S. Carollina).

Fig. 28. Symptoms of Fusarium wilt on sweet potato in the field (USDA, US Vegetable Laboratory, Charleston, S. Carollina).

Ceratocystis fimbriata

Black rot

Distribution

Ceratocystis fimbriata (*CMI Descriptions of Pathogenic Fungi and Bacteria* No. 141) occurs in North and Central America, parts of South America, West, Central and Southern Africa, India, Southeast Asia, Japan and the Pacific (CMI Distribution Map 91).

Symptoms

Black rot is an important field and storage disease of sweet potato. The fungus infects the underground parts of the plant. Infection may extend a short way up the stems causing blackening above the surface of the soil giving rise to the common name 'black shank'. The first symptoms are small, slightly sunken black spots on the underground parts of the plant. The spots enlarge and may cover almost the whole tuber. Fungal spores are often produced in the centre of these areas of infection. If infected tubers are stored the rot spreads rapidly causing heavy losses. Infected sweet potatoes have an unpleasant taste when cooked. Plants produced from infected tubers often have the disease.

Development and spread

The fungus persists in the soil between sweet potato crops. The sources of inoculum are soil and infected tubers. The disease is spread in the same way as stem rot. Plants may become infected in the seedbed or in the field. The disease spreads particularly quickly in stored sweet potatoes.

Control

The best method of control is to use disease-free planting stock. Healthy seed sweet potatoes can be grown from cuttings, which will be free from the disease as only underground parts are affected.

Cuttings should be planted in soil free from the black rot pathogen to ensure the production of disease-free roots for field planting. As the fungus persists in the soil sweet potatoes should only be grown once in every 2–3 years. Sorting out clean tubers from infected ones is not a successful method of obtaining healthy stock as many small lesions are not seen.

Sweet potatoes should not be manured with dung from animals that have eaten black rot infected tubers as it may contain spores of the fungus.

Some varieties of sweet potato have moderate field resistance to the disease.

Tubers to be used for planting can be surface sterilised with fungicides. Thiabendazole has shown promise in trials. When inoculated potatoes were dipped in a suspension of 3.32 g a.i./*l* thiabendazole development of the disease was prevented. In another trial, control given by thiabendazole and benomyl dips was greatly improved by raising the temperature of the dips to 44–54°C. Ferbam dips at 14°C were ineffective but good black rot control was obtained with dips at 44 or 54°C.

Monilochaetes infuscans

Scurf

Distribution

Australia, Brazil, China, Hawaii, Japan, Rhodesia, Sierra Leone, and USA (CMI Distribution Map 246).

Symptoms

Infection with *M. infuscans* results in the production of black spots of various shapes and sizes all over the tubers. The skin is not usually broken over these spots. The infection is only shallow and the spots can easily be scraped off. The presence of scurf does not markedly affect the nutritive value of the tubers, or the total production, however the market value is reduced considerably. Infected tubers become shrivelled due to loss of moisture when stored.

Development and spread

The fungus overseasons in stored tubers and in decayed plant material in the field. Infected seed potatoes are the main source of inoculum. Stems produced from these carry the infection to the field at transplanting. Soils which are heavy or rich in organic matter and moist conditions favour the disease.

Control

The use of clean planting stock is the most important method of control for scurf. Cuttings grown on scurf-free soil are used to produce healthy potatoes which are then planted in a clean nursery bed to produce sprouts for field planting. Visual sorting of clean roots is an inadequate method of control. Sweet potatoes should be grown in a 2 to 3 year rotation except on heavy soils and soils with a high organic matter level, which should be avoided.

The benzimidazole fungicides benomyl and thiabendazole have been shown to be highly effective for the control of scurf when used as dip treatments of seed tubers or roots of transplants. Rates used experimentally were 1.08 g a.i./*l* thiabendazole and 8.98 g a.i./*l* benomyl. Heated dips (49–50°C) were found to be more effective than tap temperature ones (18°C). Varietal differences in the effectiveness of fungicidal dips were related to the depth of tissue penetration by the fungus in the different varieties used in the trials. Maneb also gives effective control when used as a dip.

Plenodomus destruens

Foot rot

Distribution

Argentina, Hawaii, Tanzania, USA and Zaire.

Symptoms

In the field the first symptoms of infection by *Plenodomus destruens* are small brown to black spots on the stem in the region of the soil line. The infection develops slowly at first but eventually girdles the stem and extends a few inches upwards. Soon after the infection has reached this stage the plant wilts and numerous round black pycnidia (fruiting bodies) just visible to the naked eye, appear in the infected regions. Infected plants usually die, eventually. Plants infected by *P. destruens* often do not produce tubers despite good vine production. When tubers are produced they may be affected by a brown, firm rot. Fruiting bodies appear on the surface of infected tubers giving them a pimply appearance. In storage the infection can spread to healthy damaged tubers from the diseased ones.

Development and spread

Foot rot is spread in the same way as stem rot and black rot but is not as common and does not usually cause such serious losses. However, local losses can be heavy. Initial infection usually occurs in the seedbed. The pycnidia produce vast numbers of spores which can be carried to healthy plants by insects and other vectors.

Control

The control measures are the same as for stem rot and black rot, i.e. the use of healthy seed, clean seedbeds, seed treatment, and crop rotation.

Minor fungal diseases

The following diseases have been recorded on sweet potatoes but are of little economic importance or of local economic importance only:—

Phyllosticta batatas	*Phyllosticta* leaf spot
Septoria bataticola	*Septoria* leaf spot
Albugo ipomoeae panduratae (*CMI Descriptions of Pathogenic Fungi and Bacteria* No. 459)	White rust
Phymatotrichopsis omnivora	Root rot
Alternaria tenuissima	Leaf spot

Streptomyces ipomoeae

Soil rot

Distribution

USA (CMI Distribution Map 301).

Symptoms

Symptoms of soil rot are quite distinct; a heavy infection results in dwarfed plants which only produce one or two short vines with small, pale green, thin leaves. These symptoms result from damage to the underground parts of the plant.

Lateral roots are attacked and many are destroyed, others are often distorted. The most obvious symptom is the development of pits on the tubers, the first signs of which are water-soaked dark spots.

As the spots on the tubers develop the overlying skin breaks and the necrotic tissue underneath falls out leaving the characteristic pits with jagged margins. The pits range in diameter from 5–30 mm. The tubers may become girdled by lesions but continue to grow on either side of the constricted infected area.

Development and spread

The organism which causes soil rot can survive in the soil for several years and the disease is most severe in soils that are alkaline or only slightly acid. The causal organism is spread on farm implements, by people and animals and by transport of diseased tubers. Soil rot occurs widely in the USA but is a limiting factor to sweet potato production only in Louisiana where outbreaks are locally severe. Dry seasons and poor soils are conducive to severe losses.

Control

Control of this disease is difficult and the most effective method is to prevent the transfer of infected materials (soil and tubers) to uninfected fields. Planting material should be healthy. Fields known to be infested with the disease should be used for other crops for the longest period possible. Some research work has shown that the addition of sulphur to the soil as an acidifying agent will reduce the incidence of the disease, however, the acidifying effect tends to be temporary. This method should be used with caution as the soil could easily be made too acid.

VIRUS AND MYCOPLASMA DISEASES

The subject of sweet potato virus diseases is a complex one as numerous virus or virus-like diseases of this crop have been described from all over the world but for the majority details of the etiology have yet to be explained. Symptom expression varies greatly between different cultivars which adds to the confusion of disease identification. Also many different names have been used in describing the diseases and in some cases two or more names may in fact refer to the same disease. These factors make reliable field identification extremely difficult and often impossible. All adequately studied sweet potato viruses have been shown to be transmitted by grafting. Some are also transmitted by sap inoculation but this is made difficult by the presence of strong inhibitors of infection in sweet potato sap. One group of viruses is transmitted by aphids and another by whiteflies. In South Africa nematodes are suspected vectors.

The aphid-transmitted group is usually transmitted non-persistently and a number of these diseases are consistently associated with flexuous filamentous particles typical of the potato virus Y (potyvirus) group. These viruses appear to be restricted in host range to the family Convolvulaceae and can be transmitted by sap with some difficulty. This group is widespread in many countries. Some of the diseases are known to spread rapidly in the field and are therefore a serious problem. Presently available stocks of some cultivars appear to be 100% infected with one or other of these viruses, which include russet crack, virus A, vein mosaic and feathery mottle.

Viruses in the whitefly-transmitted group are often transmitted by sap inoculation and their host range extends to many plant families. This group includes mosaic, vein clearing and yellow dwarf (which are regarded by some workers as the same or very similar entities); virus B and mild mottle, which both occur in East Africa, and are also probably identical to each other.

Notes on the symptoms, transmission and causal agents of some sweet potato virus diseases are presented in Table 9. This, of necessity, presents an incomplete picture as causal agents of some diseases have either not been described, not been isolated or not been positively implicated in the etiology. However, these details need not concern those more interested in the practical aspects of crop protection in sweet potatoes.

Viruses can be a serious constraint to sweet potato production. Losses of up to 50% have been reported from South Africa. It is therefore important to be able to recognise virus infection and to know what to do about it.

Aphid-transmitted virus diseases

Various disease symptoms are associated with viruses transmitted by aphids. Russet crack symptoms have been reported from the USA and Tonga with up to 75% incidence. The main symptom is transverse cracking of the tubers. Mild infections cause brown to greyish spots or fine transverse bands with cracking, whereas with more severe infections cracking is deeper and lesions may encircle the tubers. Grey-black

TABLE 9. SWEET POTATO VIRUS DISEASES*

Symptoms	Transmission	Virus particles	Host	Area where the work was carried out	Author†
Russet crack – transverse dark lesions with cracking may encircle root. Discolouration due to secondary fungal infection. Grey spots/streaks may occur on vines or petioles	*Myzus persicae*		*I. batatas*	USA	Steinbauer and Kushman (1971)
Sweet potato **russet crack**		Pinwheel inclusions, bundle inclusions, filamentous particles	*I. batatas* *I. setosa*	USA	Lawson, Hearon and Smith (1971)
Russet crack also feathery mottle	Graft transmission *M. persicae*	Flexuous rod-shaped particle 800 nm. Member of the potyvirus group (russet crack is regarded as a strain of feathery mottle)	*I. batatas*	USA	Cambell, Hall and Mielinis (1974)
Feathery mottle complex					
i. yellow dwarf	Whitefly *Trialeurodes abutilonea*	Not described	*I. batatas*	USA	Hildebrand (1960)
ii. internal cork	Aphids } *Myzus persicae, Macrosiphum solanifolii* and others				
iii. leaf spot	Aphids }				
Internal cork – yellow spots on younger leaves. Dark brown to black hard, corky spots in tubers	Aphids	Not described		USA	Steinbauer and Kushman (1971)
Chlorotic leaf spot – always also occurs when internal cork is present in tubers	Not mentioned (but see Hildebrand, 1960)	Not described			

Internal cork		Polyhedral particles 45–47 nm	*I. batatas*		Salama *et al.* (1966)
Vein mosaic – general chlorosis, reduction in size, short internodes, distorted leaves, vein clearing, mosaic	Mechanical, sap and *Myzus persicae*	Flexuous rods 761 × 13 nm	*I. batatas* and other Convolvulaceae	Argentina	Nome (1973)
Variable – mottling and stunting or symptomless **sweet potato mild mottle**	Grafting, sap inoculation, whitefly *Bemisia tabaci*	Straight filamentous particles 800–950 nm	*I. batatas*	East Africa	Hollings, Stone and Bock (1976)
Yellow dwarf	Whiteflies *Trialeurodes abutilonea* and *B. tabaci*	Not described	*I. batatas*	USA	Steinbauer and Kushman (1971)
Sweet potato vein clearing; sweet potato chlorotic stunt – both separately latent in *I. batatas* cultivars tested in glasshouse conditions, but produce symptoms on *I. setosa*. 2 agents together produce severe symptoms on *I. batatas* (varying with cultivar) and on *I. setosa*	*M. persicae*, *Aphis gossypii; B. tabaci*	Filamentous rod 850 nm – non persistent Not detected, probably a virus	*I. batatas* *I. setosa*	Nigeria	Schaefers and Terry (1976) Schaefers and Terry (1976)
Virus A – vary with cultivar; chlorotic spots; mosaic may be latent; tends to be mild	*M. persicae*, grafting	Not described	*I. batatas*	East Africa	Sheffield (1957)
Virus B – vary with cultivar – stunting, mottling, distortion; more severe; has different strains	*B. tabaci* and sap	Not described			Sheffield (1957)
Vein clearing also mosaic, chlorosis and stunting (has different strains)	*B. tabaci*, grafting but not sap	Not described	*I. batatas*	Israel	Loebenstein and Harpaz (1960)
Ring spot conspicuous ring spots, stunting, chlorosis	*Myzus persicae*, grafting but not sap	Not described			
Leaf pucker	Not identified	Possibly a virus			

* Mycoplasma diseases not included.
† A more complete list of references can be found in the bibliography.

spots or streaks may occur on petioles and vines. Feathery mottle symptoms are frequently associated with russet crack which is now regarded as caused by a strain of the feathery mottle virus. The virus causing russet crack is serologically related to potyviruses from sweet potatoes from South Africa, East Africa, New Zealand, Nigeria and USA. It may also be related to the causal agent of internal cork.

Feathery mottle symptoms have been reported from the USA, East and South Africa. Leaves appear mottled and feathering may be observed on older leaves and in some varieties leaf distortion occurs. In later stages the infected plants become dwarfed and leaves are chlorotic and spotted. Few tubers are produced and these are small and worthless. Feathery mottle was thought to be a complex of internal cork, chlorotic leaf spot and yellow dwarf, however it has more recently been redefined as, 'an aphid-transmitted, stylet-borne virus of sweet potato that may consist of numerous strains displaying symptomatological differences but encompassing vein clearing, vein banding or feathering, chlorotic spots or ring spots'. The whitefly-transmitted yellow dwarf virus probably always occurs in addition to a strain of feathery mottle virus in field-grown plants.

Internal cork symptoms are dark brown to black, hard, corky spots in the tubers. Yellow spots develop on the leaves when they are about one month old but become less distinct as the leaves grow older. Later in the season yellow spots developing on new growth may have purple coloured borders. Corky lesions usually only develop extensively after storage at 15°C or more for several weeks to six months. Symptoms have been reported from the USA and Tonga. The causal agent of internal cork has not been conclusively identified but it has been reported to be due to two uncharacterised viruses by one author and to be associated with a polyhedral virus by another (this has not yet been confirmed). One hypothesis is that the tuber and foliar symptoms of internal cork are caused by a strain or strains of feathery mottle virus. Other published evidence indicates that the causal agent of internal cork is related to that of russet crack.

Chlorotic leaf spot symptoms are observed in conjunction with internal cork but the foliar symptoms may occur alone. The causal agent is considered to be a strain of feathery mosaic virus.

Vein mosaic disease has been reported from Argentina where it is spreading and causing considerable crop damage. Symptoms are a general chlorosis and reduction in size with a shortening of the internodes. Leaves are distorted and there is marked vein clearing. A diffuse mosaic is observed between the main veins and tubers are fewer and smaller in size. The causal virus has been observed by electron microscopy and is transmitted mechanically, by sap and by the aphid *Myzus persicae*. Despite some similarities this virus could not be positively associated with any of those described previously.

Symptoms caused by Sheffield's 'virus A' varied with the cultivar and included chlorotic spots and mottle. The causal agent was not described.

Investigations in Nigeria showed that a virus transmitted by the aphids *Myzus persicae* and *Aphis gossypii* and to a lesser extent by *A. citricola* was latent in sweet potato but produced marked vein clearing and leaf twisting in the test plant *Ipomoea setosa* (Figs. 29 and 30). These symptoms were only transient, disappearing after about four weeks.

When the causal agent of vein clearing and the whitefly-transmitted causal agent of chlorotic stunt occur together in sweet potato symptoms of vein clearing, leaf strapping, puckering, chlorosis and stunting occur. Symptom expression varies with the cultivar. On *I. setosa* dual infection causes severe symptoms of stunting and chlorosis followed by necrosis of older leaves. Further investigation on the relationship of these diseases to other sweet potato virus and virus-like diseases, such as feathery mottle, is needed.

Figs. 29 and 30. Vein clearing symptoms in sweet potato (*Ipomoea batatas*) (left) and in the indicator plant *I. setosa* (right). (E. Terry)

An aphid (*Myzus persicae*) transmitted virus causing conspicuous ring spots, stunting and necrosis has been reported from Israel. When this ring spot disease occured with whitefly-transmitted vein clearing, necrotic lesions were produced on the leaves. Attempts to transmit the causal agent of ring spot by sap inoculation were unsuccessful but it was readily graft-transmissable.

Whitefly-transmitted diseases

Mild mottle (*CMI/AAB Descriptions of Plant Viruses* No. 162) occurs in East Africa. Symptom expression varies between cultivars and tuber yield can be greatly reduced. Slight mottling, chlorotic leaf spots, veinal flecks, chlorosis, necrotic lesions and stunting have been observed. Some cultivars, particularly those with purple pigment, were found to be more susceptible than others and some were either resistant or immune. The causal virus is transmitted by *Bemisia tabaci* and has elongated filamentous particles and is probably the same as Sheffield's 'virus B'. It is serologically unrelated to russet crack or other aphid-borne viruses.

Sheffield's 'virus B', also reported from East Africa, causes symptoms of stunting, mottling and distortion which vary with the cultivar.

The causal agent of yellow dwarf is transmitted by *Bemisia tabaci* and *Trialeurodes abutilonea.* The disease, which is recorded in the USA, causes stunting of the whole plant and no useful tubers are produced. In the field the causal agent of yellow dwarf probably always occurs in conjunction with strains of feathery mottle virus.

In Nigeria, under experimental conditions, the agent which induces chlorotic stunt did not produce symptoms on sweet potato but produced mild chlorosis and severe stunting on *I. setosa.* Chlorotic stunt was transmitted by the whitefly *Bemisia tabaci.* No causal agent was detected by electron microscopy but it is thought to be a virus. As stated above, when the causal agents of chlorotic stunt and vein clearing occur together severe symptoms are caused on both sweet potatoes and the indicator plant *I. setosa.*

Work from Israel describes a whitefly (*Bemisia tabaci*) transmitted vein clearing virus. This virus caused vein clearing followed by mosaic, stunting and chlorosis. Strains of the virus, causing different symptoms were discovered. The causal agent of vein clearing was transmitted by grafting but not by sap inoculation. It was successfully transmitted to other *Ipomoea* spp.

In Georgia in the southern USA *Trialeurodes abutilonea* has been identified as a vector of a mosaic type of infection.

Hardcore

An internal hardening of the tuber tissue is noticeable after cooking. A virus was suspected to be the causal agent but chilling during storage has now been found to be responsible.

Witches' broom (little leaf)

Witches' broom disease is caused by a mycoplasma-like organism which is transmitted by leafhoppers (*Orosius ryukyuenis*) and grafting but not by *Myzus persicae* or *Cicadella viridis.* It has been reported from Taiwan and the Pacific. Mycoplasma-like bodies 200–2500 nm have been observed in the phloem of infected plants. Plants infected with witches' broom have shortened internodes giving them a bushy appearance.

Axillary buds are elongated with small leaves. Rosette, stunting, yellowing mosaic and necrosis symptoms have also been observed. Tubers are fewer and of poorer quality.

Chemical control of this disease is difficult as only tetracycline and related chemicals are effective. Heat therapy and meristem tip culture have been used to produce pathogen-free plants. Trials have identified varieties which appear to be field resistant.

The use of healthy planting material and eradication of diseased material are the most promising practical control measures at present.

It should be stressed here that the field symptoms resulting from virus infections are so variable that they are very unreliable for diagnosis. The main tests for more accurate diagnosis are graft transmission to an indicator plant (usually *Ipomoea setosa*), insect and sap transmission, serology and examination by electron microscopy.

Control

Virus infected plants should be rogued to prevent further spread of the disease. Only healthy planting material should be used. Tip cuttings may be less affected than those further down the vine. Use of healthy plant material selected on visual characteristics only is not necessarily an effective control measure as latent infections may be present. The most effective answer to this problem is to use planting material from which the virus has been eliminated by heat treatment or meristem tip culture. Heat treatment has been able to rid plants of internal cork, yellow dwarf and chlorotic leaf spot but not russet crack.

Mild mottle and some of the potyviruses have been successfully eliminated by heat treatment and meristem tip culture. Internal cork has also been eliminated by meristem tip culture.

A system for eliminating viruses and providing farmers with virus-free material has been developed and is currently in use in South Africa and could be adapted to the needs and facilities of many other countries (Fig. 31).

Sweet potato plants are grown for four to five weeks at a constant temperature of 38°C (higher temperatures may also be found to be suitable). The 0.5 mm tip cuttings are taken and after surface sterilisation they are rooted on a suitably modified growth medium in test tubes under aseptic conditions. (Work on virus indexing in Puerto Rico has shown that by using meristematic tips 0.4–0.8 mm long the need for preliminary heat treatment was avoided. In New Zealand a combination of 12 weeks heat treatment and the use of 2 mm tips was successful in eliminating the causal agent of vein clearing.) The plantlets are kept in growth cabinets for about five months and then transferred to a mist propagation unit.

When the plants are well established they are virus indexed using the wild *Ipomoea setosa* as a test plant. This is done by grafting a two leaf shoot of the sweet potato onto *I. setosa* and checking for the development of virus symptoms. The scion is encouraged to root as well as to unite with *I. setosa*. *I. setosa* shows virus symptoms much more easily than sweet potato. Indexing has to be negative three times over a year's growth before a sweet potato plant is chosen as a mother plant. If all the tests are negative the healthy material is separated from the *I. setosa* and grown on as a mother plant. Mother plants are grown in glasshouses for the cool season and encouraged to produce vines.

The mother plants are sprayed weekly with nicotine sulphate to control insect vectors. Vines are grown in screenhouses away from other sweet potato areas. Cuttings are

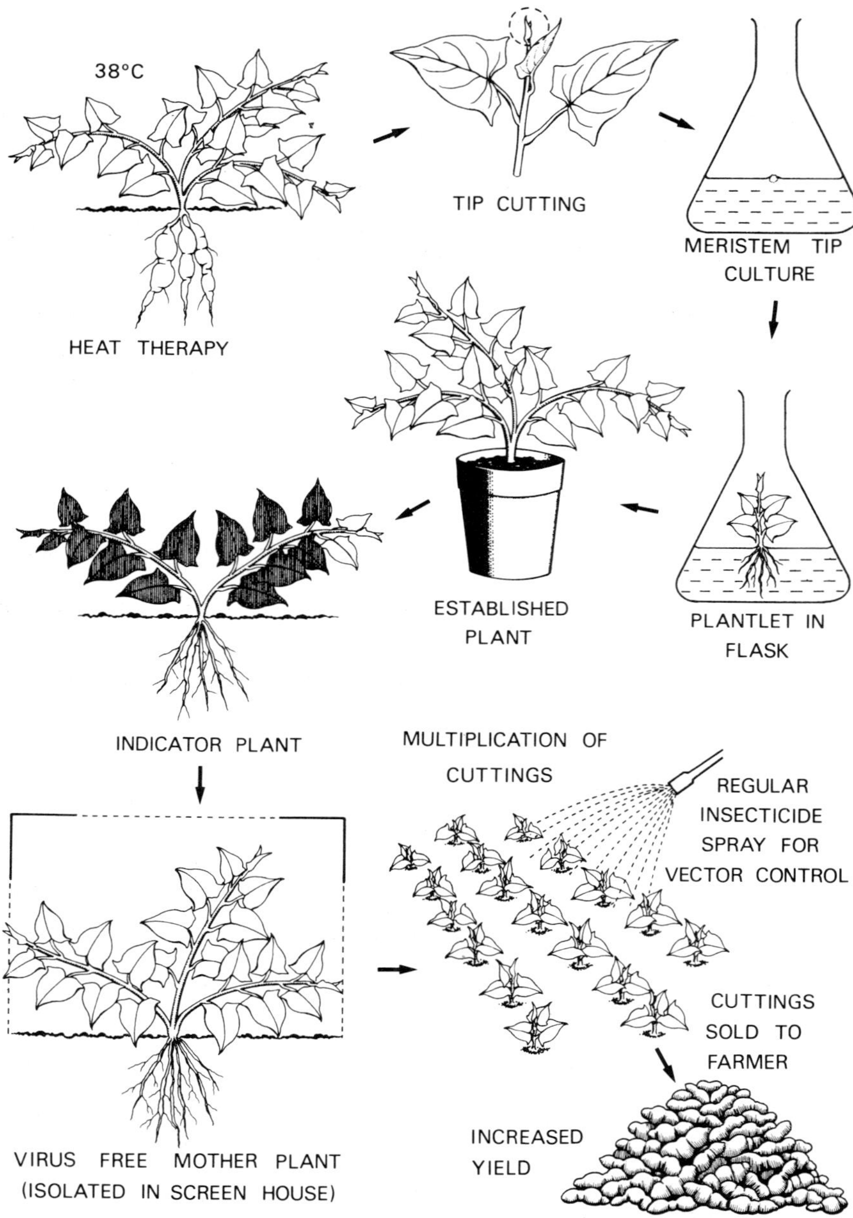

Fig. 31. Diagram showing the production of virus-free planting material by heat treatment and meristem tip culture. (H.R.B.)

taken and planted in a glasshouse for the cool season. In the next season cuttings go to farmers for multiplication and distribution. In South Africa the process of producing virus-free cuttings for distribution takes at least two years.

In some cases virus-free plants can also be obtained from seed but these are not suitable for commercial purposes as segregation may result in types different from the parent being produced. Plants produced may however be valuable for research purposes. Sweet potato is grafted onto other *Ipomoea* spp. to induce flowering.

Virus-free material should be regularly sprayed to control vectors. Weekly applications of dimethoate are recommended. Virus-free sweet potatoes should be separated by at least 400 m from other sweet potato fields which could be a source of infection.

Ideally virus-free vines would be produced by a research station and then passed on to a specialist grower who would multiply the material for sale to the commercial sweet potato farmers. This arrangement would be similar to those already existing in many countries for true seed production.

Bibliography

ALCONERO, R., SANTIAGO, A. G., MORALES, F. and RODRIGUEZ, F. (1975). Meristem tip culture and virus indexing of sweet potatoes. *Phytopathology* 65(7): 769–773.

ANON. (1976). *Short specialised training course on sweet potato production technology.* Central Tuber Crops Research Institute, Trivandrum.

BALDWIN, R. E. (1972). Control of sweet potato scurf. *Phytopathology* 62(6): 667.

CAMPBELL, R. N., HALL, D. H. and MIELINIS, N. M. (1974). Etiology of sweet potato russet crack disease. *Phytopathology* 64(2): 210–218.

DAINES, R. H. (1972). The control of black rot of sweet potatoes by the use of fungicide dips at various temperatures. *Phytopathology* 61: 1145–1146.

DAINES, R. H. (1972). Effects of fungicide dips at various temperatures on the occurrence of sweet potato scurf. *Plant Disease Reporter* 56(2): 122–126.

DAINES, R. H., CEPONIS, M. J. and HAMMOND, D. F. (1974). Relationships of chilling to development of hardcore of sweet potatoes. *Phytopathology* 64: 1459–1462.

DAINES, R. H., HAMMOND, D. F., HAARD, N. F. and CEPONIS, M. J. (1976). Hardcore development in sweet potatoes, a response to chilling and its remission as influenced by cultivar, curing temperatures, and time and duration of chilling. *Phytopathology* 66: 582–587.

GIRARDEAU, J. H. (1958). The sweet potato whitefly *Bemisia inconspicua* Q., as a vector of a sweet potato mosaic in south Georgia. *Plant Disease Reporter* 42: 819.

GLASSHOUSE CROPS RESEARCH INSTITUTE (1972). *1971 Annual Report.* p. 120 and pp. 130–135.

GRAHAM, K. M. (1971). *Plant diseases of Fiji.* Overseas Research Publication No. 17. pp. 250. Ministry of Overseas Development, London.

HAMMOND, D. F., DAINES, R. H. and CORBETT, M. K. (1974). The association of cellular inclusions and virus like particles with 'hardcore' of fleshy sweet potato roots. *Plant Disease Reporter* 58(1): 17–20.

HAMMOND, D. and DAINES, R. H. (1975). A virus in centennial sweet potatoes as the agent of russet crack and internal cork. *Proceedings of the American Phytopathological Society* 1974 1: 150.

HILDEBRAND, E. M. (1960). The feathery mottle virus complex of sweet potato. *Phytopathology* 50: 751–757.

HILDEBRAND, E. M. (1961). Relations between whitefly and sweet potato tissue in transmission of yellow dwarf virus. *Science, New York* 133: 282–284.

HILDEBRAND, E. M. (1964). Heat treatment for eliminating internal cork viruses from sweet potato plants. *Plants Disease Reporter* 48: 356–358.

HILDEBRAND, E. M. (1967). Russet crack—a menace to sweet potato industry. I. Heat therapy and symptomatology. *Phytopathology* 57: 183–187.

HILDEBRAND, E. M. (1968). Russet crack—a menace to the sweet potato industry. II. Natural transmission. *Plant Disease Reporter* 52: 967–971.

HOLLINGS, M., STONE, O. M. and BOCK, K. R. (1976). Purification and properties of sweet potato mild mottle, a white-fly borne virus from sweet potato (*Ipomoea batatas*) in East Africa. *Annals of Applied Biology* 82(3): 511–528.

HOLLINGS, M. and STONE, O. M. (1977). Virology-sweet potato (*Ipomoea batatas*). *Report of the Glasshouse Crops Research Institute for 1976.*

JOUBERT, T. G. la G., KLESSER, P. J. and NEL, D. D. (1974). *IV. The cultivation of sweet potatoes. D.1. The value of virus-indexed material.* pp. 3. Department of Agricultural Technical Services, Pretoria, South Africa.

KAHN, R. P., LAWSON, R. H., MONROE, R. L. and HEARON, S. (1972). Sweet potato little-leaf (witches' broom) associated with a mycoplasma-like organism. *Phytopathology* 62 (8): 903–909.

LAWSON, R. H., HEARON, S. S. and SMITH, F. F. (1971). Development of pinwheel inclusions associated with sweet potato russet crack virus. *Virology* 46 (2): 453–463.

LOEBENSTEIN, G. and HARPAZ, I. (1960). Virus diseases of sweet potatoes in Israel. *Phytopathology* 50: 100–104.

MACDONALD, A. S. (1963). Sweet potatoes with particular reference to the tropics. *Field Crop Abstracts* 16 (4): 219–225.

MARTIN, W. J. (1972). Effect of thiabendazole on development of stem rot of sweet potato. *Phytopathology* 62 (7): 776.

MARTIN, W. J. (1972). Further evaluation of thiabendazole as a sweet potato 'seed' treatment fungicide. *Plant Disease Reporter* 56 (3): 219–223.

NIELSEN, L. W. (1960). Elimination of internal cork virus by culturing apical meristems of infected sweet potatoes. *Phytopathology* 50: 840–841.

NOME, S. F. (1973). Sweet potato vein mosaic virus in Argentina. *Phytopathologische Zeitschrift* 77 (1): 44–54.

OVER DE LINDEN, A. J. and ELLIOTT, R. F. (1971). Virus infection in *Ipomoea batatas* and a method for its elimination. *New Zealand Journal of Agricultural Research* 14: 720–724.

SALAMA, F. M., LYMAN, C. M. and WHITEHOUSE, V. G. (1966). Isolation of the internal cork virus from sweet potato leaves having the purple ring structure by use of differential filtration. *Phytopathologische Zeitschrift* 57: 89–95.

SCHAEFERS, G. A. and TERRY, E. R. (1976). Insect transmission of sweet potato disease agents in Nigeria. *Phytopathology* 66 (5): 642–645.

SHEFFIELD, F. M. L. (1957). Virus diseases of sweet potato in East Africa. I. Identification of the viruses and their insect vectors. *Phytopathology* 47: 582–590.

SHEFFIELD, F. M. L. (1958). Virus diseases of sweet potato in East Africa. II. Transmission to alternative hosts. *Phytopathology* 48: 1–6.

SMITH, S. N. and SNYDER, W. C. (1971). Relationship of inoculum density and soil types to severity of *Fusarium oxysporum batatas* wilt of sweet potato. *Phytopathology* 61 (9): 1049–1051.

SO. I. Y. (1973). Studies on the mycoplasmic witches'-broom of sweet potato in Korea (1) symptoms of pathogen. *Korean Journal of Microbiology* (1973) 11 (1): 19–30.

STEINBAUER, C. E., and KUSHMAN, L. J. (1971). *Sweet potato culture and diseases.* Agriculture Handbook No. 388. pp. 74. ARS, US Department of Agriculture.

TERRY, E. R. (1976). A review of the major pests and diseases of sweet potato with special reference to plant exchange and introduction. *African Journal of Plant Protection* 1(1): 117–133.

NEMATODES*

***Meloidogyne incognita, Meloidogyne* spp.**

Root-knot nematodes

Distribution

Probably widespread in the majority of sweet potato growing areas. They have been reported from Brazil, French West Indies, Ghana, Ivory Coast, Japan, Nigeria, Puerto Rico, Peru, South Africa, Trinidad and USA.

Symptoms

Meloidogyne incognita is the major root-knot species damaging sweet potato, but the crop is also attacked by other *Meloidogyne* species, *M. javanica, M. hapla* and *M. arenaria.*

The infective second stage larvae of *M. incognita* invade the root tissues at the tips of young roots in the region of tissue differentiation, through the loose ruptured cells in enlarging roots where lateral roots emerge, and via the broken surface of cracks in the root. They migrate through the tissues and select feeding sites in the stele of root tips, and in the bud primordia, secondary cortex, cambial zones and vascular tissues of enlarging roots. Feeding initiates the formation of 'giant cells' around the nematode's head, also cell enlargement and cell multiplication in the tissues surrounding the nematodes as they mature. In most infections, feeding also stimulates the formation of abnormal xylem tissue. Eggs are laid in a matrix either deep in root tissues or extruded onto the surface in primary roots. In the older, enlarged roots, nematodes and egg masses become surrounded by layers of suberised cells. Only a few nematodes are found in the centres of large roots.

Nematode infections in primary roots cause swelling (knotting) of the entire root, and heavy infections can inhibit apical growth. The most obvious symptoms of damage on enlarged roots are longitudinal cracking and general rough appearance of the skin (Fig. 32). In many cases, the lenticels are noticeably enlarged. Constriction of enlarging roots can occur in some sweet potato varieties infected with root-knot. Both the roots and tops of infected plants are reduced.

Control

Root-knot nematodes are spread in the seed roots from infested sweet potato and, for complete control, nematodes have to be eliminated both from the field soil and seed roots. There is some evidence suggesting that clean cuttings of seed roots from the apex are less prone to root-knot damage when planted out than cuttings from the basal end.

---

*This chapter was written by Dr J. Bridge, ODM Liaison Officer in plant nematology.

Fig. 32. Two tubers of cv. Impala (left), highly resistant to *Meloidogyne* spp. and two tubers of cv. Hartbees (right), very susceptible, grown in soil infested with *Meloidogyne* spp. (Horticultural Research Institute, Pretoria)

Chemical soil treatment has proved to be a successful and economic means of controlling root-knot nematodes on sweet potatoes in South Africa, the USA and Brazil. The following are examples of the nematicides and rates which have been shown to increase the quality and yield of root-knot susceptible varieties:

1. EDB applied by chisel injection in rows 20 cm deep at a rate of 23.5 *l* a.i./ha;
2. D-D applied by chisel injection 20 cm deep in rows at 93.5 *l*/ha;
3. Fensulfothion applied as granules in a 30 cm band at 3.4 kg a.i./ha;
4. Carbofuran applied as granules 1 to 4 days before planting at a rate of 6.7 kg a.i./ha;
5. Prophos as granules applied broadcast at 6.7 kg a.i./ha, or in a 30 cm band at 3.4–4.5 kg a.i./ha two weeks before planting.

Total yield and number of marketable roots can also be increased by treating transplant beds with nematicides, e.g. prophos and aldicarb granules spread on the soil surface at a rate of 1.04 g/30 cm^2

Physical control methods using both hot water and hot air treatments are effective in eliminating root-knot and other nematodes from infested roots. Immersion in hot water at 45°C for 30 h, 46.7°C for 65 min, and 50°C for 3–5 min gives complete control of nematodes without seriously impairing the viability of roots. Hot air treatments of roots, 5–8 cm in diameter, at 50°C for 4–8 hours eliminates most, if not all, root-knot nematodes without significantly affecting their keeping qualities. However, some varieties of sweet potato are less heat tolerant than others.

A number of sweet potato varieties are resistant to one or more species of *Meloidogyne*. Examples are: Heartogold, Nemagold, Redmar, Jersey types, the breeding line W-13, Koganesenga and Norin No. 5; variety Jasper is moderately resistant and Painter shows some tolerance.

Rotylenchulus reniformis

Reniform nematode

Distribution

Reported on sweet potatoes from Cuba, French West Indies, Ghana, Jamaica, Nigeria, Puerto Rica, Senegal, Trinidad and USA, but is certainly more widespread than this.

Symptoms

Rotylenchulus reniformis is normally a semi-endoparasite of roots with a swollen portion of the female body protruding from the root surface when mature; eggs are laid in an egg-sac which is extruded from the female into the soil. Eggs hatch and develop into immature females which invade the roots. Normally *R. reniformis* does not give rise to gall formation, but on sweet potato it has been found that the nematodes become surrounded by galls formed from host tissues 9 days after they infect the roots.

R. reniformis is associated with chlorosis of the foliage and a reduction in yield and size of swollen roots. The most serious injury occurs when nematode population levels are high at planting time and there is a seasonal increase of the nematodes.

Control

Chemical control of *R. reniformis* in the field can greatly increase yields of marketable and total roots. The nematicides and rates used against root-knot nematodes will also be effective in controlling *R. reniformis*. Good control of *R. reniformis* and increases in sweet potato grade have been achieved by soil injection of D-D at 93.5 *l*/ha 2–3 weeks before planting; granular application in 30 cm band of fensulfothion at 5.6 kg a.i./ha; prophos at 3.4–6.7 kg a.i./ha and aldicarb at 5.6 kg a.i./ha.

Hot water treatment of rootlets by immersing in water at 50°C for 3–5 min completely eliminates *R. reniformis*. Other pre-plant heat treatments used against root-knot nematodes will also control *R. reniformis*.

The reniform nematode is able to reproduce on a wide range of crop and weed hosts thus making crop rotation difficult as a method to control the nematodes. However, in Trinidad, *R. reniformis* soil populations have been reduced by fallow and in rotation with susceptible crops. The sweet potato variety Goldrush is less susceptible to *R. reniformis* than other varieties, but is highly susceptible to root-knot nematodes.

Other nematodes

Many other plant parasitic nematodes have been found associated with sweet potato roots and could be adversely affecting growth when present in large populations. They include: *Belonolaimus* sp., *Criconemoides* spp., *Ditylenchus destructor, Helicotylenchus* spp., *Hoplolaimus* sp., *Longidorus* sp., *Macroposthonia* sp., *Paratylenchus* sp., *Pratylenchus* spp., *Radopholus similis, Scutellonema* spp., *Tylenchorhynchus* spp., *Tylenchus* spp., *Xiphinema* spp.

Bibliography

AVERRE, C. W., NIELSEN, L. W. and BARKER, K. R. (1974). Organic pesticides for controlling *Meloidogyne incognita* on *Ipomoea batatas. Phytopathology* 64: 767.

AYALA, A. and RAMIREZ, C. T. (1964). Host-range, distribution and bibliography of the nematode *Rotylenchulus reniformis* with special reference to Puerto Rico. *Journal of Agriculture and University of Puerto Rico* 48: 140–161.

BIRCHFIELD, W. and MARTIN, W. J. (1968). Evaluation of nematicides for controlling reniform nematodes on sweet potatoes. *Plant Disease Reporter* 52: 127–131.

BOOCK, O. J. and LORDELLO, L. G. E. (1963). Nova contribuičao ao estudo do contrôle de nematóides em cultivas de batata doce. *Revista de Agricultura, Sao Paulo* 38: 31–38.

BRAITHWAITE, C. W. D. (1972). Preliminary studies of plant parasitic nematodes associated with selected crops at the University of the West Indies. *Plant Disease Reporter* 56: 1077–1079.

BRAITHWAITE, C. W. D. (1974). Effect of crop sequence and fallow on populations of *Rotylenchulus reniformis* in fumigated and untreated soil. *Plant Disease Reporter* 58: 259–261.

BRAITHWAITE, C. W. D. and DUNCAN, E. J. (1974). Development and histopathology of *Rotylenchulus reniformis* in sweet potato roots. *Tropical Agriculture* 51: 437–441.

BURK, E. F. and TENNYSON, G. (1941). Hot water treatment for control of nematodes in sweet potato seed roots. *Proceedings American Society of Horticultural Science* 39: 299–302.

CORDNER, H. B., STRUBLE, F. B. and MORRISON, L. (1954). Breeding sweet potatoes for resistance to the root-knot nematode. *Plant Disease Reporter,* Supplement No. 277, pp. 92–93.

DICKERSON, O. J., ROBBINS, R. T. and GREIG, J. K. (1969). Sweet potato transplants protected from root-knot nematode infection by chemical treatment of transplant beds. *Journal of Nematology* 1: 284–285.

GIAMALVA, M. J., MARTIN, W. J. and HERNANDEZ, T. P. (1963). Sweet potato varietal reaction to species and races of root-knot nematodes. *Phytopathology* 53: 1187–1189.

JOHNSON, A. W., HARMON, S. A. and CHALFANT, R. B. (1974). Influence of organic pesticides on nematode and insect damage and on yield and grade of sweet potato. *Plant Disease Reporter* 58: 239–243.

JOUBERT, T. G. la G. and DE BRUYN, J. (1967). Impala is also eelworm resistant. *Farming in South Africa* 43(4): 13.

KONDO, T. (1972). [Varying rate of reproduction of *Meloidogyne incognita* (Kofoid & White) Chitwood, in different varieties of sweet potato.] *Japanese Journal of Nematology* 2: 16–19.

KRUSBERG, L. R. and NIELSEN, L. W. (1958). Pathogenesis of root-knot nematodes to the Porto Rico variety of sweet potato. *Phytopathology* 48: 30–39.

MARTIN, W. J. (1970). Elimination of root-knot and reniform nematodes and scurf infections from rootlets of sweet potato plants by hot water treatment. *Plant Disease Reporter* 54: 1056–1058.

NIELSEN, L. W. and SASSER, J. N. (1959). Control of root-knot nematodes affecting Porto Rico sweet potatoes. *Phytopathology* 49: 135–140.

ORTIZ, R. E., OCAÑA LOPEZ, C. and VARGAS CHIAPPE, L. (1971). Effect of slip type on the histological differentiation, early ripening, production and susceptibility to nematodes of *Ipomoea batatas* cv. 'Japanese'. *Nematropica* 1: 38–43.

THOMASON, I. J., VAN GUNDY, S. D. and McKINNEY, H. E. (1960). Thermotherapy for root-knot nematodes, *Meloidogyne* spp. of sweet potato and tarragon propagating stocks. *Plant Disease Reporter* 44: 354–358.

VAN BLOMMESTEIN, J. A. and STOFBERG, F. J. (1957). What causes cracks in sweet potatoes? Vexed question solved. *Farming in South Africa* 32(10): 42–43, 45.

INSECT PESTS

Introduction

The most common pest of sweet potato is *Cylas formicarius* which reduces tuber yield and quality, although there are simple cultural techniques that reduce the losses due to this pest. Young plants are susceptible to foliage feeders, such as sweet potato hornworm and it may be necessary to treat the unestablished plants with insecticide. Good crop hygiene and crop rotation are particularly important in the production of sweet potato as a number of pests may be carried over from one season to the next in the soil or on weeds. *Cylas* spp. and *Alcidodes* spp. continue to damage the tubers in storage.

LEAF FEEDERS

Brachmia spp. (Lepidoptera: Gelechiidae)

Leaf folder

Distribution

Brachmia convolvuli: Canary Is., Mauritius, Comoro Is., Southeast Asia and Java. *Brachmia macroscopa* has been recorded in India.

Symptoms

Cream and black larvae spin webs close to the main veins of the leaves, on the underside, while older larger larvae spin webs across the upper surface of the leaf and draw the two sides together. The leaves eventually dry up, die and drop off.

Tabidia aculealis (in Malaya) and *Pilocrocis* sp. in the West Indies cause similar damage.

Life cycle

Brachmia convolvuli eggs are laid on the lower surface of the leaves, and hatch after 3 days (in the laboratory). There are five larval instars. The mature larva is 11–12 mm long and cream coloured with dorsal black bands on segments 8–10 of the abdomen; the head, prothoracic shield and all of abdominal segment 5 are black. The larvae pupate for 7 days in the folded leaf. Adult *B. convolvuli* are 4 mm long with greyish brown forewings, marked with white, and white hindwings fringed with grey. The entire life cycle, egg to adult takes 24 days.

Control

Sprays of 0.2% carbaryl or 0.025% fenitrothion are recommended in India for *Brachmia convolvuli,* in the West Indies 1 kg a.i./ha of carbaryl or trichlorphon in high or low volume sprays are recommended against *Pilocrocis* sp.

Agrius convolvuli (Lepidoptera: Sphingidae)

Sweet potato hornworm, convolvulus hawkmoth

Distribution

Most of Africa, Australia, Bangladesh, Burma, South China, southern Europe, India, Indonesia, Iran, Malaysia, New Zealand, Pacific Islands and Papua New Guinea.

Symptoms

Leaves are eaten by hornworms (Fig. 33). The taro hornworm (see p. 200) may also be found on sweet potato.

Life cycle

Spherical eggs (1 mm diameter) are laid singly on the host plant, usually on the upper leaf surface. The larvae, which concentrate on the young shoots, are variable in colour from green to brown and distinctively patterned. There are five larval instars and larvae mature after 3–4 weeks and reach a length of about 95 mm. Pupation takes place in the soil and may take 3 weeks or several months depending upon the climate. The adult hawkmoths (wingspan 80–120 mm) are grey with black lines on the wings; the abdomen is marked laterally with pink bands. They feed upon nectar in flowers.

Control

Cultural

Ploughing the land between crops to expose the pupae reduces infestation. Hand picking of the larvae may be quite effective in small areas.

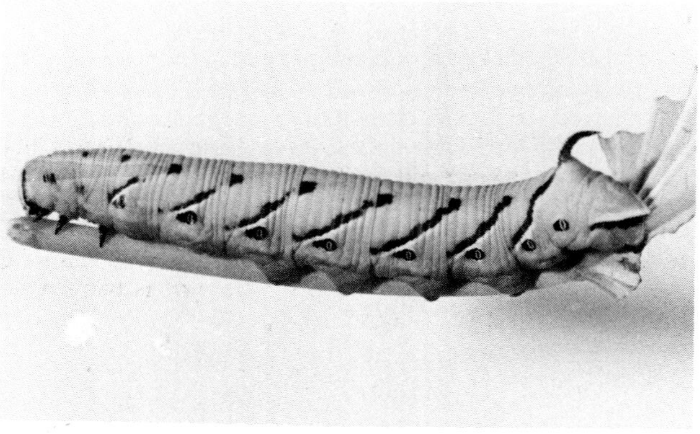

Fig. 33 *Agrius convolvuli,* sweet potato hornworm, larva of the convolvulus hawkmoth. (G. Swaine (1971) *Agricultural Zoology in Fiji*)

Chemical

A 0.2% solution of carbaryl, a 0.25% solution of DDT in a high volume spray, 0.05% endosulfan or trichlorphon (7 day waiting period before harvest) 100 g a.i. in 100 *l* water should be applied when the larvae are first seen, early instars can be controlled more easily than fifth instars. Particular care should be taken to spray the young leaves.

Diacrisia obliqua **(Lepidoptera: Arctiidae)**

Bihar hairy caterpillar

Distribution

South and East Asia.

Symptoms

Numerous larvae covered in yellow hairs skeletonise the leaves, entire vines may be defoliated.

Life cycle

Eggs are laid on the leaves in rows or clumps and after 4–9 days the larvae emerge. Mature larvae are 30–40 mm long with seven orange bands across the body on which grow tufts of yellow hairs. At first the larvae feed in groups but when they are about a week old they disperse to new feeding sites. The larval period lasts 28–34 days. Pupation takes place in the soil in reddish-brown cocoons and may take 51 days.

Control

Chemical

BHC dust (10% a.i.) at 20–22 kg/ha is recommended in India for the control of the pest when the larvae are still gregarious. Endosulfan at 0.05% 1000 *l*/ha and fenitrothion at 0.05% are also suggested. Diazinon, or dichlorvos at the manufacturers recommended rate and 0.2% carbaryl will also control these larvae, and similar leaf-feeding Lepidoptera such as *Euchromia polymena* and *Swinhoea vegetus* reported in India.

Bedellia **spp. (Lepidoptera: Lyonetiidae)**

Found nr Wabag.

Sweet potato leaf miners

Distribution

Bedellia ipomoeae is recorded from Fiji.

Bedellia somnulentella: Papua New Guinea and Queensland.
There are a number of *Bedellia* spp. attacking sweet potatoes in Africa.

Symptoms

Narrow mines widening to blotches appear in the leaves. The plants have a withered appearance and growth is reduced.

Life cycle

Eggs are laid on the leaf. The emerging larva is distinctly segmented with a rather pointed head and abdomen. Larvae may be red or yellowish grey in colour, after making several mines they pupate on a web of silken threads among the leaves. When moving outside the mines the larvae travel by a looping movement. The life cycle can be completed within 4 weeks. Adult *Bedellia* spp. are small (3 mm long with a wingspan of 8 mm) tan to grey moths.

Control

Chemical

Generally *Bedellia* spp. are controlled by predators and parasites but these may be killed by foliar sprays of insecticide. A serious outbreak can cut down the effective leaf surface and lead to reduced yields of tubers, Table 10 sets out the chemical control methods advised in the Pacific area.

TABLE 10. RECOMMENDATIONS FOR CHEMICAL CONTROL OF *BEDELLIA* SPP.*

Insecticide	Formulation	Active ingredient	Application
Carbaryl	w.p.	1 kg/ha	In 13.6 *l* water
Chlorfen-vinphos	dust	0.6–0.9 kg/ha	In 13.6 *l* water
Diazinon	e.c.	55 ml/ha	In 13.6 *l* water
Dimethoate	e.c.	14 ml/ha	In 13.6 *l* water
Endosulfan	—	70 ml/ha	In 1000 *l* water
Malathion	w.p.	14 g/ha	In 13.6 *l* water
Trichlorphon	w.p.	100 g/ha	In 100 *l* water. 7 day minimum period before harvest

* Spray twice with an interval of one week.

Aspidomorpha spp. (Coleoptera: Chrysomelidae)
Conchyloctaenia punctata

Tortoise beetles

Distribution

Conchyloctaenia punctata (large spotted tortoise beetle) South Africa.

Aspidomorpha adhearens, A. australasiae, A. miliaris, A. quadriradiata and *A. socia* are recorded attacking sweet potato in Papua New Guinea.

Aspidomorpha miliaris is also described with *Metriona circumdata* as a pest of sweet potato in India.

Aspidomorpha dissentanea is recorded from Angola, Cameroon and Guinea.

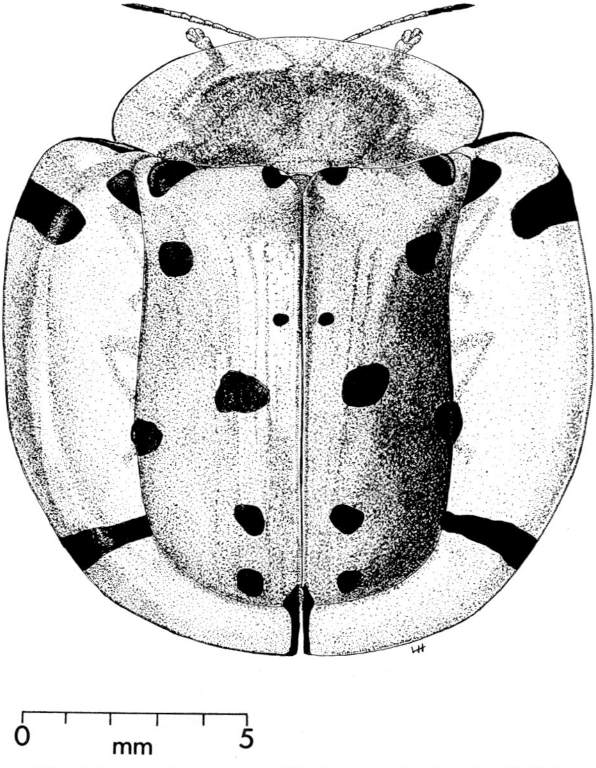

Fig. 34. *Aspidomorpha miliaris,* a tortoise beetle. (L.H.)

Symptoms

Adults and larvae feed on the leaves making large round holes and eventually skeletonising the leaves.

Life cycle

The eggs are laid on the underside of the sweet potato leaves (or other Convolvulaceae) in batches cemented to the leaf. The larvae are flattened, spiny and usually carry excreta and the skins of earlier instars on anal projections. Adult tortoise beetles (sub-family Cassidinae) are broadly oval in shape (Fig. 34) hence the common name. They are bright, sometimes patterned or metallic in colour.

Control

Cultural

Weeds of the family Convolvulaceae should be controlled as tortoise beetles can breed on them.

Chemical

Carbaryl at 1000 *l*/ha 0.2% is recommended in India, DDT 0.05–0.25% or 1 kg a.i./ha carbaryl in Papua New Guinea and endosulfan at 500–750 g a.i./500 *l* water/ha elsewhere.

TUBER FEEDERS

Cylas spp. (Coleoptera: Curculionidae)

Distribution

Cylas formicarius is pantropical (CIE Distribution Map 278).

Cylas puncticollis is confined to Africa, it has been recorded from most countries south of the Sahara (CIE Distribution Map 279).

Cylas brunneus: West Africa.

Symptoms

Tubers and vines are tunnelled by small white, legless larvae. The tunnels may be partially filled with frass. Ant-like weevils (Fig. 35), either having glossy blue black elytra, reddish-brown legs and thorax and a black head (*C. formicarius*) or entirely shiny black (*C. puncticollis* or *C. brunneus*) may be found on the leaves or in the tunnels. Yield, storage life and plant vigour are reduced.

Life cycle

The total length of development is 4–5 weeks. Eggs are laid in holes in the vines or tubers and hatching takes place within a week if the temperature is between 20 and 30°C. The larvae tunnel in the vines and tubers for about 14 days. When fully developed (fourth instar) the larvae are 3 mm in diameter. Pupation takes place inside the plant and lasts about one week.

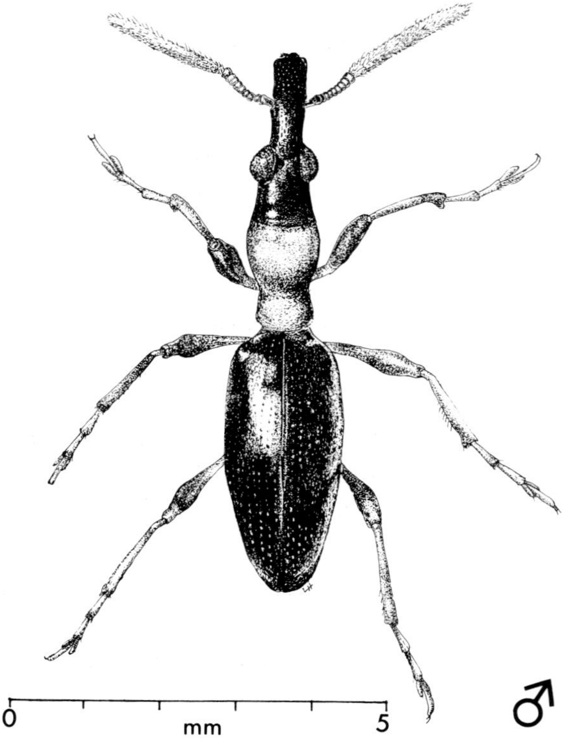

Fig. 35. *Cylas formicarius,* sweet potato weevil. (L.H.)

Control

Cultural

Sweet potatoes should be planted only in fields that have had no *Cylas* infestation within the last 12 months and preferably more than 1 km away from any infested land, and only *Cylas* free planting material should be used. Resistant varieties should be chosen, if available; deep rooting varieties with long necks between the tubers and the stems being less susceptible because the adult weevil cannot burrow downwards more than 1 cm. In some areas early maturing varieties escape serious damage. Earthing-up the plants, particularly those varieties with a tendency to push out of the ground reduces attack. After harvest the field should be cleared of trash, which should be burned or fed to stock.

Chemical

See Table 11 for recommended chemical control methods.

TABLE 11. RECOMMENDATIONS FOR CHEMICAL CONTROL OF *CYLAS* SPP.

Insecticide	Formulation	Active ingredient	Application
Azinphos-methyl	Follow manufacturers recommendations		
γ -BHC	dust	1.2 kg/ha	Apply at 14 day intervals when the weevil is observed on the foliage. 30 day minimum interval before harvest.
γ -BHC	—	0.1%	Soil drench fortnightly
Carbaryl		0.1%	Three times at 4 weekly intervals.
Carbofuran	gran.	1400 g/ha	A single application to the soil around the vine cutting at planting time
Carbophenothion	—	0.035%	—
Chlorfenvinphos	dust	600–900 g/ha	30 day minimum interval before harvest
Chlorfenvinphos	e.c.	240 g/ha	—
Diazinon	Follow manufacturer's recommendations		Dip for planting material or foliar spray
	—	0.3%	—
Dieldrin		1 kg/ha	4 times at 2 week intervals beginning 40 days after planting. 5 week minimum interval before harvest
Disulfoton	—	1.8 kg/ha	4 weeks after planting
Endosulfan	e.c.	2.25 kg/ha	Apply at planting time. Do not feed the foliage to animals

Insecticide	Formulation	Active ingredient	Application
Endosulfan	gran.	2.25 kg/ha	Incorporate into the soil at planting time. Do not feed the foliage to animals
Fenitrothion	—	0.05%	
Fenthion	—	0.1%	3 times at 4 week intervals
Malathion	e.c.	28.5 ml	In 13.6 *l* water, dip for cuttings or foliar spray
Mephosfolan	gran.	1.4 kg/ha	In the soil around the cutting at planting
Methomyl		1 kg/ha	4 times at 2 week intervals beginning 40 days after planting
Methyl-demeton	—	0.05%	Spray at the beginning of the second, third and fourth month of growth
Monocrotophos	—	0.05%	4 times at monthly intervals
Parathion	Follow manufacturer's recommendations		
Phorate	gran.	1.4 kg/ha	In the soil around the cutting at planting
	—	2 kg/ha	4 weeks after planting
Phosphamidon	—	0.05%	Spray at the beginning of the second, third and fourth month of growth

Alcidodes spp. (Coleoptera: Curculionidae)

Distribution

Alcidodes dentipes (striped sweet potato weevil) has been recorded from tropical Africa.

Alcidodes erroneus has been recorded on sweet potato in Kenya.

Alcidodes fabricii has been recorded attacking sweet potato in India.

Symptoms

Alcidodes dentipes (Fig. 36) and *A. erroneus* attack both stems and tubers. Adults girdle the stems causing wilting, while larvae bore into stems and tubers making tunnels slightly larger than those of *Cylas* larvae and causing galls to form on the stems.

Alcidodes fabricii has been recorded breeding inside the seed capsules of sweet potato.

Life cycle

The life cycle of the weevils is thought to be similar to that of *Cylas*. *Alcidodes dentipes* is about 14 mm long and brown with four white streaks along its elytra. *Alcidodes erroneus* is similar in size but brown-black in colour with an irregular white patch on each elytron.

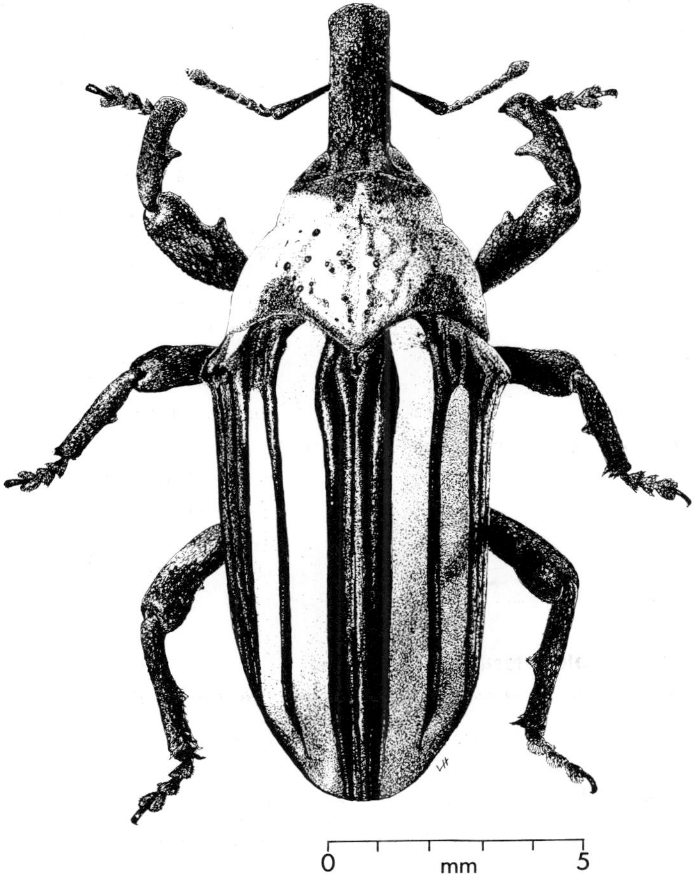

Fig. 36. *Alcidodes dentipes,* the striped sweet potato weevil. (L.H.)

Alcidodes fabricii is reddish-brown with four longitudinal white stripes on the elytra. The eggs are laid at the base of the fruit, the larvae bore into the fruit and feed on the seeds. Pupation takes place within the fruit.

Control

The methods outlined for *Cylas* should control *Alcidodes* spp. if necessary.

Euscepes postfasciatus (Coleoptera: Curculionidae)

Scarabee weevil, West Indian sweet potato weevil

Distribution

Central America, Bermuda, Brazil, Cook Islands, Fiji, Futuna Is, French Polynesia, Guam, Guyana, New Caledonia, Ryuku Is, Tonga, southern USA and West Indies.

Symptoms

Larvae are seen feeding in the tubers. This weevil is particularly important in dry localities.

Fig. 37. *Euscepes postfasciatus,* scarabee weevil or West Indian sweet potato weevil. (L.H.)

Life cycle

Eggs are laid in the underground parts of the plant where they hatch within 7 days. Larvae are about 5 mm long, legless, white with a yellowish head and reddish-brown mandibles when full grown. Development to pupation takes 25–28 days; pupation takes place in the tunnel made by the larva. The adult emerges from the pupal skin after one week but remains in the pupal cell until the integument hardens. Adults (Fig. 37) are greyish-brown weevils with a white mark near the apex of the body.

Control

Cultural

Crop rotation, use of insect-free cuttings, earthing up the plants and cleaning up crop trash after harvest will reduce the risk of infestation.

Chemical

Dipping cuttings in diazinon or malathion emulsifiable concentrate made up in water to the strength recommended by the manufacturers enables the farmer to use infested cuttings for propagation, although they should preferably be avoided. Chlorfenvinphos, diazinon, dieldrin and disulfoton are recommended in the West Indies and Hawaii for the control of both *Cylas* and *Euscepes*—see Table 11 for details.

Megastes grandalis (Lepidoptera: Pyralidae)

Sweet potato pyralid moth, stem borer

Distribution

Brazil and Trinidad.

Symptoms

Plants are stunted and may shed their leaves. Frass can be seen on the soil around the crown of the plant. Cuttings may die due to larvae feeding inside them. Harvested tubers, when cut open, are found to be riddled with clean tunnels. Even a slight infestation can be important because it may precipitate rot.

Life cycle

Green, plump oval eggs 1.2–1.5 mm in diameter are laid in groups beside the veins on the underside of the leaf or in the axils of the petioles. As they mature they turn purple.

The larvae are white with brown head and markings when they emerge but after the second instar the body is pinkish. The larvae begin to feed near the egg as soon as they hatch and a short time later they move closer to the soil and bore into the stem and on into the tuber. The frass produced by tunnelling is usually pushed out through the first entry hole. When it has fed in the tuber for 5–7 weeks the full grown larva spins a cocoon not far from the hole above ground through which the adult will emerge. The pupal period lasts 13–16 days. The moths mate and begin to lay eggs the night after they emerge. Sweet potato is the only known host of this insect.

Control

Chemical

The chemicals in Table 12 are recommended for the control of *M. grandalis* in the West Indies.

TABLE 12. RECOMMENDATIONS FOR CHEMICAL CONTROL OF *MEGASTES GRANDALIS*

Insecticide	Formulation	Active ingredient	Application
Carbaryl	—	950 g/ha	In 900 *l* water (high vol) or 225 *l* water (low volume) at 3 weekly intervals
Chlorfen-vinphos	e.c.	300 ml/ha	In 900 or 225 *l* water at 3 weekly intervals. 30 day minimum period before harvest
Dieldrin	—	0.5%	Dip cuttings before planting
Malathion	—	1.4 kg/ha	
Methomyl	w.p.	500–950 g/ha	In 900 or 225 *l* water at 3 weekly intervals. 7 day minimum period before harvest if feeding foliage to stock

Omphisa anastomosalis (Lepidoptera: Pyraustidae)

Sweet potato stem or vine borer

Distribution

Generally widespread in Southeast Asia; China, India, Japan and Sri Lanka.

Symptoms

Stems wilt and die and black-headed, yellowish-white larvae, marked with brown tubercles can be found boring in the stems. Frass can be found close to the attacked stems. These larvae may also bore into the tubers.

Life cycle

The flattened yellow, oval eggs are laid in groups of 2 or 3, usually on the stem and occasionally on the leaves. The larvae emerge after an incubation of 5–8 days and eventually reach 30 mm in length. They tunnel in the stem for 28–50 days and pupate in a thin cocoon within the tunnel, close to the tubers. Adults emerge after about 14 days. The moth is 15 mm long with a wingspan of 33 mm, with a reddish-brown body and reddish-brown markings on the wings, which are white.

Control

Cultural

Crop hygiene and crop rotation reduce infestation.

Chemical

Losses due to *Omphisa anastomosalis* have been estimated as 400 kg of tubers/ha in Malaysia but it may not be as important elsewhere. In Malaysia carbaryl at 0.1% a.i. or 0.1% a.i. BHC sprayed at fortnightly intervals is recommended.

Synanthedon dasysceles **(Lepidoptera: Sesiidae)**
Synanthedon leptosceles

Clearwing moth

Distribution

Kenya.

Symptoms

The main stem develops galls at ground level, the plant is unhealthy and the yield is reduced. Tubers are tunnelled by white larvae with three pairs of thoracic legs and five abdominal prolegs.

Life cycle

Comparatively little is known of the life cycle. Adults are day flying and lay batches of 70–100 eggs, probably on the vines. It is thought that the larvae burrow into the vines on hatching and tunnel downwards, frequently reaching the tubers. Usually pupation takes place in the main stem of the plant just above ground level. When the adult has emerged the brown pupal cocoon and pupal case may be seen protruding from the stem.

Control

The recommendations for *Cylas* control (Table 11) will reduce the incidence of this pest.

Bibliography

ANON. (1976). Insects of traditional vegetables. Topic 11. *Entomology Conference, Papua New Guinea, 1976.* Unpublished report, Department of Primary Industry.

ANON. (1976). Vegetable pests. Topic 13. *Entomology Conference, Papua New Guinea, 1976.* Unpublished report, Department of Primary Industry.

BERG, H. C. van den and RAPPARD, C. E. (1972). The sweet potato pest, are its days numbered. *Farming in South Africa* 46(7): 14–15.

BRADLEY, J. D. (1967). Two new species of clearwing moths (Lepidoptera, Sesiidae) associated with sweet potato (*Ipomoea batatas*) in East Africa. *Bulletin of Entomological Research* 58: 47–53.

BUTANI, D. K. and VARMA, S. (1976). Pests of vegetables and their control: Sweet potato. *Pesticides* 10(2): 36–38.

CADEILHAN GIRANDET, L. (1969). Contribution a l'etude d'une Chrysomelidae *Aspidomorpha dissentanea* nuisible a la patate douce. *Boletim Instituto de investigaceo cientifica de Angola* 5(1): 5–28.

COMSTOCK, J. A. (1966). Lepidoptera of American Samoa with particular reference to biology and ecology. *Pacific Insects Monograph* 11: 1–74.

COWLAND, J. W. (1926). Notes on the sweet potato pyralid moth, *Megastes grandalis* Guen. *Bulletin of Entomological Research* 16: 369–372.

CTCRI (1971). *Annual Report 1970.* pp. 78. Central Tuber Crops Research Institute, Trivandrum.

CTCRI (1972). *Annual Report 1971.* pp. 102. Central Tuber Crops Research Institute, Trivandrum.

EULITZ, E. G. (1974). *Die biologie van die patatkalander.* Msc. thesis, University of Pretoria.

FENNAH, R. G. (1947). *The insect pests of food crops in the Lesser Antilles.* pp. 207. Department of Agriculture, Antigua, BWI.

HILL, D. (1975). *Agricultural insect pests of the tropics and their control.* pp. 516. Cambridge University Press, Cambridge, UK.

HO, T. H. (1970). Studies on some major pests of sweet potatoes and their control. *Malaysian Agricultural Journal* 47(4): 437–452.

HORTICULTURAL RESEARCH INSTITUTE (1975). *The cultivation of vegetables in South Africa; IV The cultivation of sweet potatoes; A1 The production of sweet potatoes.* pp. 8. Department of Agricultural Technical Services, Pretoria.

INGRAM, W. R. (1967). Chemical control of sweet potato weevil in Uganda. *East African Agricultural and Forestry Journal* 33: 163–165.

JAYARAMAIAH, M. (1975). Studies on the chemical control of the sweet potato weevil, *Cylas formicarius* (Fabricius) Olivier (Coleoptera: Curculionidae). *Mysore Journal of Agricultural Sciences* 9(2): 307–313.

KHURANA, A. D., VERMA, A. N. and BHANOT, J. P. (1974). Chemical control of leaf feeding caterpillar *Brachmia macroscopa* Meyr. on sweet potato. *Haryana Agricultural University Journal of Research* 4(2): 173–175.

LEVER, R. J. A. W. (1933). Some insect pests of local economic plants *British Solomon Islands Protectorate Gazette* 1(1): 3–5.

LOWE, S. B. and WILSON, L. A. (1972). Preliminary evidence for the existence of differential susceptibility to *Megastes grandalis* (Guen) infestation in West Indian sweet potato cultivars. *Tropical Agriculture (Trinidad)* 49(4): 361–362.

MATHEW, K. P. and ABRAHAM, C. C. (1972). Occurrence of *Swinhoea vegeta* Swinh. (Lepidoptera: Noctuidae) as a pest of sweet potato in Kerala. *Agricultural Research Journal Kerala* 10(2): 193.

O'CONNOR, B. A. (1969). *Exotic plant pests and diseases.* pp. 424. South Pacific Commission, Noumea, New Caledonia.

PARASRAM, S. (1973). *Control of insect pests of some food crops in the Caribbean. Extension Bulletin* No. 7. pp. 29. University of the West Indies.

PILLAI, K. S. (undated). *Pests of sweet potato and their control.* Central Tuber Crops Research Institute, Trivandrum.

PILLAI, K. S. and MAGOON, M. L. (1969). Studies on chemical control measures for sweet potato weevil *Cylas formicarius* Fb. *Indian Journal of Horticulture* 26(3–4): 202–208.

PILLAI, K. S., RAO, Y. R. V. J. and MANDAL, R. C. (1973). Sweet potato pests can be controlled. *Indian Farming* 22(10): 22–23 and 46.

SANCHEZ, F. F. and MEDINA, J. R. (1975). Screening of insecticides for sweet potato weevil control. *Plant Protection News* (Philippines) IV(2): 23.

SINGH, S. R. (1973). Identification of resistance to and insecticidal control of sweet potato weevil, *Cylas puncticollis. Proceedings of the Third International Symposium on Tropical Root Crops,* Nigeria.

SINGH, S. R. (undated). *Common root crop pests of Fiji and their control.* pp. 10. Ministry of Agriculture Fisheries and Forests, Fiji.

SMEE, L. (1965). Insect pests of sweet potato and taro in the territory of Papua and New Guinea: their habits and control. *Papua and New Guinea Agricultural Journal* 17(3): 99–101.

SUBRAMANIAN, T. R. (1954). Sweet potato as a new host plant for the weevil, *Alcidodes fabricii* Fabricius. *Current Science* 7: 234–235.

SUBRAMANIAN, T. R., SUBBA RAO, P. V. and GOPALSWAMY, N. (1973). Evaluation of certain new insecticides in the control of the sweet potato weevil, *Cylas formicarius* Fb. *Madras Agricultural Journal* 60(7): 621.

TANAKA, J. S. and SEKIOKA, T. T. (1976). Sweet potato in Hawaii. *Proceedings of the Fourth International Symposium on Tropical Root Crops, Cali, Colombia,* 1976. p. 50. International Development Research Centre, Ottawa, Canada.

VISALAKSHI, A. and JOHNSON, J. (1968). Studies on a leaf feeding caterpillar of sweet potato *Brachmia convolvuli* Wlsm. (Gelechiidae: Lepidoptera). *Agricultural Research Journal Kerala* 6(1): 30–32.

WHEATLEY, P. E. (1961). The insect pests of agriculture in the coast province of Kenya; III Sweet potato. *East African Agricultural and Forestry Journal* 26: 228–229.

WILSON, L. A. and LOWE, S. B. (1973). Development of supplementary translocatory tissues in intact and *Megastes* infested tuber stalks of sweet potato (*Ipomoea batatas* L.Lam). *The Journal of Horticultural Science* 48(3): 223–226.

WOOD, I. J. L. (1976). Sweet potato growing in Queensland. *Queensland Agricultural Journal* 102(6): 553–566.

CASSAVA

DISEASES

Introduction

In the past cassava has been thought of as a hardy crop with no significant disease problems. However, it has been shown that diseases can cause serious losses. Where cassava has been grown on research stations yields of up to 40 t/ha have been obtained—showing a considerable improvement over the world average of 10 t/ha. Under traditional farming systems yields have been more than doubled by using disease free planting material and improving cultural practices.

As cassava is propagated vegetatively the use of disease free planting material is the most important single factor in disease control. Cassava bacterial blight is the most important disease of cassava and the virus diseases are also a considerable threat to the crop. This chapter aims to give concise information on cassava diseases and their control. It should be noted that the distributions quoted for the diseases have been compiled from information available at the time of writing and may therefore be incomplete owing to the lack of relevant disease records.

FUNGAL DISEASES

Cercosporidium henningsii

Brown leaf spot

Distribution

Widespread in all cassava growing areas.

Symptoms

Brown leaf spot is probably one of the most widespread fungal leaf diseases of cassava. Older leaves tend to be more susceptible than younger ones and the disease tends to be more widespread and severe in crops over five months old. However, susceptibility varies between cultivars.

The disease causes characteristic brown spots (3–12 mm in diameter) on both sides of the leaves. Spots on the upper surface are angular and uniformly brown with a distinct darker border (Fig. 38). Spots on the lower surface have less distinct margins and appear greyish in the centre because of the presence of fruiting bodies of the fungus.

Fig. 38. Brown leaf spot (*Cercosporidium henningsii*). Large brown lesions with distinct borders. (Lozano and Booth (1974), *PANS* 20(1): 30–54)

The lesions expand and become irregular and angular in shape as they are limited by the leaf margins and veins. Commonly an indefinite yellow halo appears round the lesions. As the disease develops the infected leaves turn yellow, become dry and fall off. Susceptible varieties can be severely defoliated during the rainy season.

Development and spread

Warm, wet conditions favour the development and spread of *C. henningsii.* Conidia of the fungus on older or fallen leaves are spread by rain splash and wind to younger leaves and to new plantings. The conidia need moisture to germinate on the leaf surface. The fungus penetrates through the stomata and grows through the intercellular spaces. The fungus survives during the dry season in old lesions, often on fallen leaves.

Control

In many areas losses caused by brown leaf spot have never been found to merit the recommendation of control measures. In some areas of Colombia where the rainfall is over 120 mm losses of over 20% have been recorded. Where the disease is prevalent the best control measure is to plant resistant varieties. Work is in progress on further development of these in many areas, including Africa and South America.

Measures designed to reduce the humidity within the stand of cassava, such as wider spacing and the use of varieties with a low leaf area index are also recommended. The severity of the disease can be reduced if planting is timed so that the most susceptible growth stage (over 5 months old) does not coincide with the wet season. Copper fungicides, thiophanate, benomyl and other fungicides will control the disease but their use is rarely economic.

Phaeoramularia manihotis

White leaf spot

Distribution

Occurs commonly in humid but cooler cassava growing areas.

Symptoms

White leaf spot occurs less commonly than brown leaf spot and is not so important. The lesions are smaller (1–7 mm) than those of brown leaf spot. The spots are circular to angular, white or yellowish brown in colour and sunken from both sides to about half the thickness of a healthy leaf. On the lower surface of the leaf the lesions often have a diffuse coloured border that appears as an irregular violet-brown line surrounded by a yellow halo (Fig. 39). The middle of the lesions may have a greyish velvety appearance owing to the presence of fruiting bodies of the fungus.

Development and spread

Similarly to *C. henningsii*, *Phaeoramularia manihotis* is spread by rain splash and wind. The fungus penetrates through the stomata and invades the tissues through the intercellular spaces. The fungus overseasons on old infected leaves.

Control

Control measures are similar to those recommended for brown leaf spot. Copper fungicides or zineb give good control but their application is rarely economic. Destruction of infected and fallen leaves is also recommended as a phytosanitary measure. Observations in the field suggest that there is some resistance to this pathogen but specific resistant varieties have yet to be developed.

Fig. 39. White leaf spot (*Phaeoramularia manihotis*). Small white lesions with distinct violet/brown border and diffuse yellow halos. (Lozano and Booth (1974), *PANS* 20(1): 30–54)

Cercospora viscosae

Cercospora leaf blight

Distribution

Warm cassava growing areas of Brazil and Colombia and also southeast Asia. Occurs as commonly as brown leaf spot.

Symptoms

This disease is characterised by large brown spots without definite borders. Individual spots may cover a fifth or more of the leaf lobe. Similarly to brown leaf spot the lesions are uniformly brown on the upper surface but on the lower surface the centres of the spots are greyish due to the presence of fruiting bodies of the fungus. The lesions are very similar to those of concentric-ring leaf spot. However, lesions of the latter usually have concentric rings on the upper surface.

Cercospora leaf blight usually occurs during the rainy season in areas where brown leaf spot is prevalent and can cause heavy defoliation in susceptible cultivars with resulting yield loss of over 30%. Plants over six months old are more severely affected.

Control

The same as for brown leaf spot.

Resistant varieties have been identified and should be planted where possible.

Phoma sp., *Phyllosticta manihobae*, *Phyllosticta manihoticola*

Concentric-ring leaf spot

Distribution

Africa, Brazil, Colombia, India and the Philippines.

Taxonomy

The causal agent of this disease has not been clearly defined and further studies on its taxonomy are needed. Several *Phyllosticta* species have been associated with the same disease symptoms but it is possible that there is a single causal agent. Recent studies indicate that the causal agent is a *Phoma* sp. but, until further studies are carried out, it is not known whether *Phyllosticta manihobae* and *Phyllosticta manihoticola* are synonymous with this or not.

Symptoms

This disease occurs in cooler cassava growing areas during the rainy season when temperatures are below 22°C. Susceptible varieties may be severely defoliated and occasionally dieback or death may occur. The disease is characterised by large brown spots with indefinite margins. Lesions are usually found at the tips or edges of the leaf lobes or along the main veins or midrib. At first the upper surface of the lesions consists of concentric rings formed by the brown pycnidia (fruiting bodies) of the fungus (Fig. 40). These rings are often absent from older lesions as the epidermis disintegrates allowing mature pycnidia to be washed off by rain so that the lesions resemble those caused by *Cercospora viscosae*. Few pycnidia are produced on the lower surface of the lesions, which appear uniformly brown.

When humidity is high greyish-brown hyphae can be observed on the surface of the lesions. On the underside of leaves veins and veinlets become necrosed in the region of the lesions. The lesions expand causing a leaf blight. Eventually the necrosis spreads to the whole leaf and petiole. Leaves then wilt and drop, sometimes resulting in extensive defoliation. Dieback of young shoots can also be caused and in extreme cases the plant may die. Diseased stems become brown and are often covered with pycnidia.

Fig. 40. *Phyllosticta* leaf spot (*Phyllosticta* spp.). Leaflets showing large brown lesions with concentric rings. Young stems showing die-back and presence of pycnidia. (Lozano and Booth (1974), *PANS* 20(1): 30–54)

Development and spread

Field observations suggest that younger upper leaves are more susceptible to the disease. The disease is most prevalent at high altitudes or in lowland areas during the rainy season, because conditions are best for spore germination (maximum spore germination occurs between 20 and 25°C). The way in which the fungus overseasons is not yet understood.

Control

No specific control measures for this disease are reported in the literature. Field resistant cultivars have been observed and these may be used in a breeding programme for resistance.

Oidium manihotis

Cassava ash disease

Distribution

Worldwide. Endemic in almost all cassava plantations.

Symptoms

The first sign of the disease is a white fungal mycelium growing over the surface of the leaf. Yellowish lesions are formed where the fungus penetrates the leaf (Fig. 41). The lesions are indefinite and contain pale brown angular spots of necrotic tissue. In some varieties the lesions remain yellow and undefined. The symptoms may be confused with damage caused by some insects and spider mites.

The disease occurs most commonly during the dry season in warmer cassava growing areas. Mature leaves appear to be the most susceptible but younger leaves of some varieties are also infected.

Fig. 41. Cassava ash disease (*Oidium manihotis*). Typical undefined yellow leaf lesions. (Lozano and Booth (1974), *PANS* 20(1): 30–54)

Control

Cassava ash disease does not appear to be of economic importance and specific control measures are not generally considered necessary. Resistance has been observed in some varieties.

Glomerella manihotis

Anthracnose

Distribution

Widespread.

Symptoms

Anthracnose tends to appear after long periods of rain. Young shoots seem to be the most susceptible. Leaf border blight and spots appear on young leaves, which become distorted. Infected tissue becomes necrotic and whole leaves may die. The disease also attacks stems, causing a dieback of young stems and the production of cankers on older stems. In the centre of the cankers pinkish areas of the fungal fruiting bodies can be observed. Anthracnose tends to disappear at the onset of the dry season. When severe, symptoms may resemble African mosaic disease because of leaf distortions.

Control

Anthracnose appears to be serious in southeast Asia, Africa and Brazil but generally it is considered to be of minor importance. However, planting material taken from infected stems is of poor quality, does not germinate well and results in weak plants and reduced yield. The best preventative measure is to use healthy planting material and to avoid planting at the height of the rains.

Sphaceloma manihoticola

Superelongation disease

Distribution

Colombia.

Symptoms

In infected plants young stems show exaggerated elongation and appear thin and weak. Diseased plants are considerably taller than healthy ones. Young shoots, petioles and leaves are frequently deformed. Young leaves are often slightly chlorotic and have many whitish spots. These deformations are associated with lens-shaped cankers which form along the midribs or veins of leaves and on the young stems and petioles (Figs. 42 and 43). The leaves are often incompletely developed and young leaves may become partially or totally necrotic. Dieback occurs frequently and there may be considerable defoliation.

Figs. 42 and 43. The superelongation disease (*Sphaceloma manihoticola*). General symptoms showing young stem and petiole elongation, leaf deformation and distortion and cankers (left). Leaf showing leaf curl symptoms, irregular white spots on the leaf lamina and cankers on the midribs, veins and petiole (right). (Lozano and Booth (1974), *PANS* 20(1): 30–54)

Development and spread

Superelongation is most severe during the rainy season and is spread by the use of infected cuttings and wind flow. Volunteer plants are a major source of infection. Conidiospores are produced in the cankers which are acervuli. Spores are believed to be spread by means of rain or wind. At the start of the dry season the infection and spread of the disease decline. The disease increases in importance at the resumption of wet conditions as free moisture is necessary for conidia to germinate. Initially the fungus grows on the epidermis and following direct penetration grows in the intercellular spaces of the epidermis and cortex. The cankers are formed when mycelial aggregates push up the cortex and cause the epidermis to rupture.

Control

Superelongation disease occurs in epidemics and early infection has been reported to result in up to 80% yield loss. Tubers are produced but they are smaller and thinner than normal. Late infection does not result in significant losses.

The use of resistant varieties is the best method of control. Field resistance to this disease occurs so resistant varieties should be planted where possible. Fungicide trials with copper hydroxide, benomyl and maneb have shown promise. The disease can be eradicated from infected cuttings by treating them with a captafol (80%) dip at 4,000 ppm a.i. for 3 min.

Uromyces spp.

Rusts

Six different species of the order Uredinales have been reported on cassava in Latin America; of these *U. manihotis* is the most widely distributed. Generally rusts are not important diseases of cassava. Typical rust pustules are formed on leaves and stems. Some species are prevalent in cool, wet conditions while others are favoured by hot, dry conditions. The mycoparasite *Darluca filum* restricts the sporulation of *Uromyces* spp. on cassava in some areas.

Stem rots

In areas where cassava is not grown continuously it is necessary to store stems for planting in the next season. Stem rot diseases can cause considerable loss of viability during this storage period.

Three distinct diseases have been recognised but in practice it is often difficult to distinguish between them as their macroscopic appearance may be similar and more than one pathogen may be present.

Glomerella stem rot

Glomerella is the most common stem rot of stored cassava cuttings. The fungus is also found on stem debris left in cassava plantations. The rot is first seen at the cut ends and then slowly spreads along the cuttings. The vascular strands develop a black discolouration which is followed by the development of surface blisters. These later burst revealing the black perithecia (fruiting bodies) of the fungus (Fig. 44). Infection probably takes place through wounds and is favoured by high humidity. The relationship between the causal agent, probably *Glomerella cingulata* (*CMI Descriptions of Pathogenic Fungi and Bacteria* No. 315), and the causal agent of anthracnose has not yet been determined.

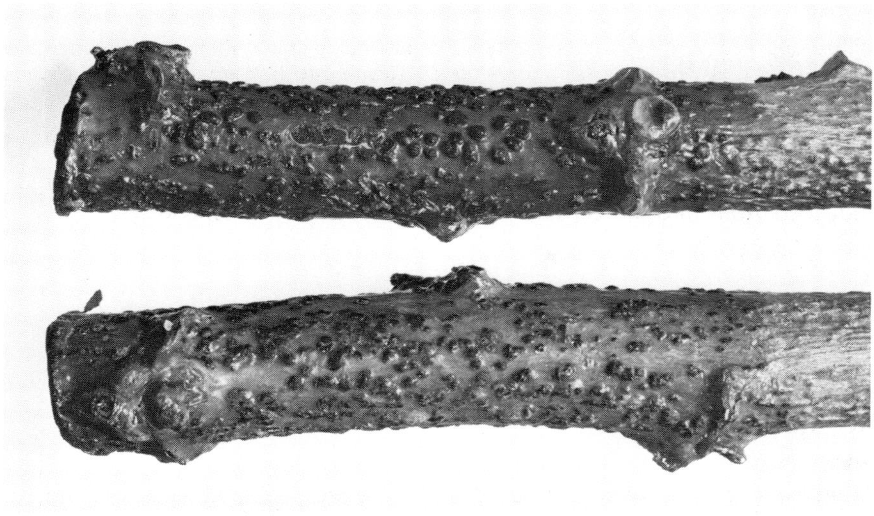

Fig. 44. Glomerella stem rot (*Glomerella cingulata*). Pieces of stem showing eruptive blisters and groups of black perithecia. (Lozano and Booth (1974), *PANS* 20(1): 30–54)

Botryodiplodia stem rot

Botryodiplodia theobromae (*CMI Descriptions of Pathogenic Fungi and Bacteria* No. 519) infects stem cuttings and crop debris but is less common than *Glomerella* rot. Vascular strands show a black discolouration and necrosis radiating from wounds. Epidermal blisters are produced under which the infected dark brown or black tissues can be seen. Masses of black pycnidia can be seen when these blisters rupture. Wounds and high humidity favour the disease.

Unidentified stem rots

Several unidentified Basidiomycetes and Ascomycetes appear on stem pieces in the field and in storage (Fig. 45). Infected stem pieces are necrotic with a slight brown to black discolouration. Sometimes white mycelium can be seen growing under the epidermis. Occasionally under humid conditions basidiocarps or ascocarps of different sizes and colours develop on badly infected cuttings.

Control

Stem cuttings should be carefully selected and handled with care to avoid wounding. Storage humidity should not be too high but care should be taken to avoid desiccation of the cuttings. The use of 'seed-treatment' fungicides has been shown to be highly

Fig. 45. Stem rot (unidentified basidiomycete). Pieces of stem showing typical basidiocarps, together with an uninfected control. (Lozano and Booth (1974), *PANS* 20(1): 30–54)

advantageous and routine treatment is recommended. Captan + carbendazim, chlorothalonil + mancozeb, mancozeb, chloroneb, quintozene and copper oxychloride are recommended (2,000 ppm a.i. in mixtures; 4,000 ppm a.i. when used alone). Mixtures provide wider control.

Minor leaf and stem diseases

Microsphaera euphorbiae	Powdery mildew
Verticillium dahliae (*CMI Descriptions of Pathogenic Fungi and Bacteria* No. 256)	Vascular wilt
Drechslera rostrata	Leaf blight
Phomopsis manihotis	Blight
Corticium salmonicolor (*CMI Description of Pathogenic Fungi and Bacteria* No. 511)	Pink disease
Periconia sp.	Leaf spot

Septobasidium sp., the cause of grey felt fungus, occurs as a parasite of scale insects on cassava. It is sometimes mistakenly thought to be a disease of cassava.

Root rot diseases

Cassava pre-harvest root rot diseases are important in areas with badly drained soils and during periods of excess rainfall or when cassava is planted after a woody crop such as coffee, rubber or forest trees. Damping-off during the early stages of plant growth and rotting of the swollen roots are caused by many pathogens. When roots of young plants are attacked, serious yield losses can result. Multiple infection occurs frequently and there is often secondary infection by weak parasites and saprophytes so that the initial causal agent cannot be identified. Symptoms of the root rot diseases therefore often appear similar.

Phytophthora drechsleri

Phytophthora root rot caused by *P. drechsleri* has been reported from both Africa and tropical America and yield losses of up to 80% have been recorded. The pathogen affects both mature and young plants causing sudden wilting and severe soft rotting of the swollen roots (Figs. 46 and 47). Diseased roots develop a brown discolouration, and disintegrate rapidly. Roots exude a foul smelling liquid when partially rotted. In severe cases plants may die. *P. erythroseptica* and *P. cryptogea* are also reported as causal agents of cassava root rot. *Phytophthora* root rot is prevalent on waterlogged soils.

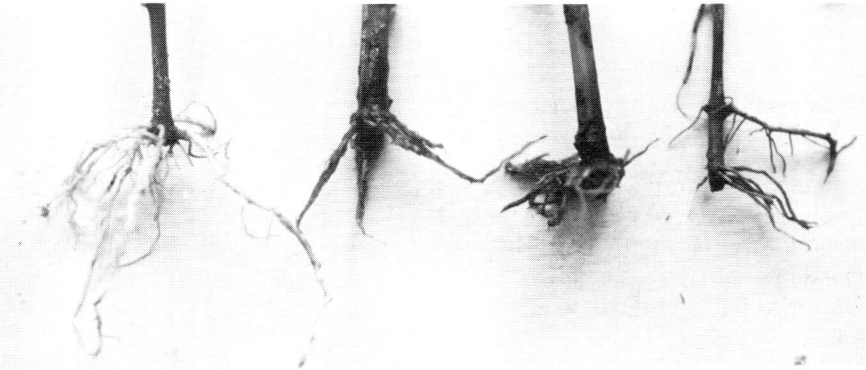

Fig. 46. *Phytophthora* root rot (*Phytophthora drechsleri*). Root rot of young plants together with an uninoculated control. (Lozano and Booth (1974), *PANS* 20(1): 30–54)

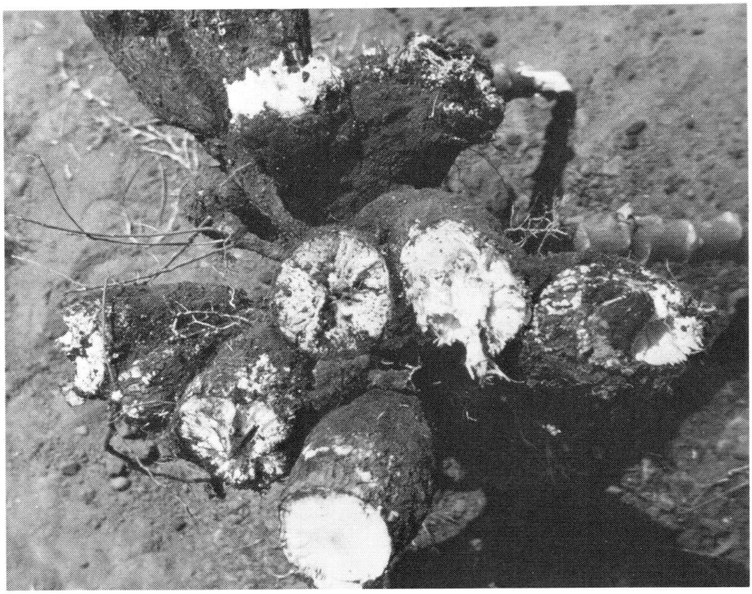

Fig. 47. *Phytophthora* root rot (*Phytophthora drechsleri*). (Lozano and Booth (1974), *PANS* 20(1): 30–54)

Rigidoporus lignosus

White root disease

This disease (*CMI Descriptions of Pathogenic Fungi and Bacteria* No. 198) is the most serious root rot disease of cassava in Africa, Malaysia, Sri Lanka and Thailand. In Latin America the disease is only of major importance in the Amazonas area of Brazil and in Mexico where forest has been cleared. The disease is most common when cassava is planted after jungle clearance or a susceptible crop such as rubber.

The fungus forms a white mycelial mat under the bark of infected swollen roots and white mycelial threads coat the exterior of infected roots (Fig. 48). Internally the roots appear dry and smell of rotting wood. Sometimes young plants wilt suddenly and lose their leaves when all the roots become necrotic. The occurrence of this disease is patchy; this is thought to be because the fungus is unable to spread any great distance through the soil and infection occurs as a result of direct contact with previously colonised roots.

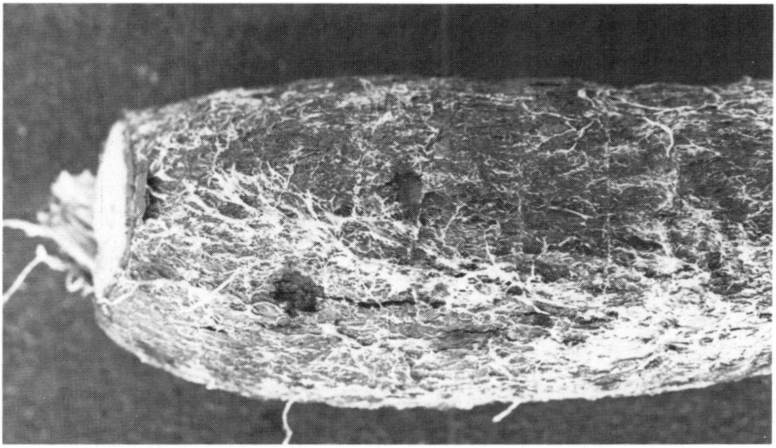

Fig. 48. White thread disease. (Booth and Wholey)

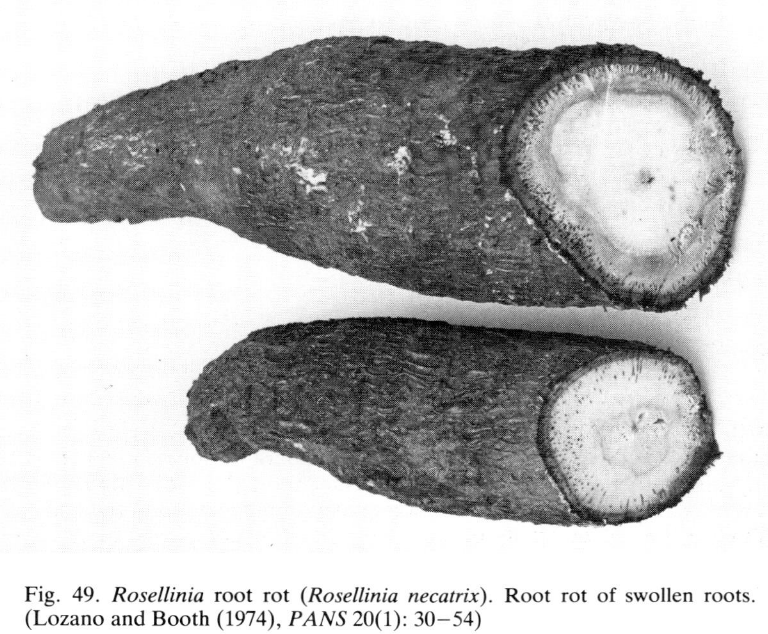

Fig. 49. *Rosellinia* root rot (*Rosellinia necatrix*). Root rot of swollen roots. (Lozano and Booth (1974), *PANS* 20(1): 30–54)

Rosellinia necatrix

Rosellinia root rot, black rot

This disease occurs in many cassava growing areas where cassava is planted after forest clearance and in soils which are high in organic matter and appears to be mainly confined to South America. *R. necatrix* (*CMI Descriptions of Pathogenic Fungi and Bacteria* No. 352) causes a characteristic black discolouration and cankers (Fig. 49). Root surfaces are initially covered with white mycelial strands or rhizomorphs which later turn black. Internal tissues are slightly discoloured and exude liquid when squeezed. Black mycelia spread through the tissues and small cavities containing mycelium may be formed. Infected roots smell of rotting wood. The disease is not reported to attack young plants.

Corticium (*Sclerotium*) *rolfsii*

Corticium root rot (*CMI Descriptions of Pathogenic Fungi and Bacteria* No. 410) occurs on roots of cassava at most stages of development. The white mycelium can be seen radiating into the soil from infected tissues. If the fungus penetrates roots through wounds a rot results causing considerable necrosis.

Frog skin disease

This disease is not common but can cause tremendous losses when it occurs. Affected roots are abnormally thickened and their skin becomes corky, giving the roots a cracked, wrinkled appearance. The causal agent of the disease has not been isolated but it can be transmitted by cuttings. Healthy planting material should be used.

Minor root rot diseases

Armillariella mellea (*CMI Descriptions of Pathogenic Fungi and Bacteria* No. 321)

Pythium sp.

Pseudophaeolus baudonii

Helicobasidium compactum

Fusarium sp.

Clitocybe tabescens

Rhizopus spp.

Rhizoctonia spp.

Aspergillus spp.

These pathogens have been recorded as causal agents of root rot of cassava but little information is available on their importance. Bacteria are also often associated with root rots.

Control

Cassava root rot diseases can sometimes be controlled by cultural practices such as good drainage, use of lighter soils, early harvest and crop rotation depending on the causal organism present. Waterlogged soils, such as the areas near drainage ditches, should be avoided. After clearing woody crops or forest, non-susceptible crops, such as cereals, should be planted to avoid infection with woody pathogens. Stumps should be completely removed when clearing new land as these may serve as a reservoir for pathogens such as *Rigidoporus lignosus*. Wide spacing tends to discourage the spread of root pathogens. Crop debris should be removed and burnt and healthy planting material should be selected. Rotation with cereals and fallowing help to lower the inoculum level in the soil.

BACTERIAL DISEASES

Xanthomonas manihotis

Cassava bacterial blight (CBB)

Distribution

Argentina, Brazil, Cameroon, Colombia, Ghana, India, Indonesia, Madagascar, Malawi, Malaysia, Mauritius, Mexico, Nicaragua, Nigeria, Panama, Philippines, Taiwan, Thailand, Togo, Uganda, Venezuela, Zaire (CMI Distribution Map 521).

Causal organism

The CBB causal organism is an aerobic, Gram-negative, non-sporeforming rod, motile bacterium with a single polar flagellum, not encapsulated and $0.5-0.6 \times 0.9-2\mu m$ in size. On sugar media colonies are whitish.

The bacterium causing CBB is at present considered to be a strain of *Xanthomonas manihotis* (*CMI Descriptions of Pathogenic Fungi and Bacteria* No. 559) because, although it differs from the type cultures of *X. manihotis,* it is close enough to be considered the same species. However, a revision of the taxonomy of this group is necessary.

The organism named *X. cassavae* which was isolated from cassava and forms yellow colonies has sometimes been considered to be a strain of *X. manihotis*. The most recent edition of Bergey's Manual groups over a hundred different pathogens, including *X. cassavae* and *X. manihotis* under *X. campestris.* It appears that there is variation in virulence and characteristics of different strains of the causal organism of CBB.

Symptoms

CBB is the most important disease of cassava and is widely distributed. It is particularly devastating during wet seasons.

Fig. 50. Cassava bacterial blight. Plant showing typical leaf wilt and dieback symptoms. (Lozano and Booth (1974), *PANS* 20(1): 30–54)

Characteristic symptoms are leaf spotting and blight, wilting, dieback, gum exudation and stem and root vascular necrosis. Primary symptoms, shown by infected planting material, are wilting of the young germinated shoots followed by dieback.

Symptoms of secondary infection are angular leaf spotting, followed by blight, defoliation, wilting and dieback (Fig. 50). At first the leaf spots are small, angular and water soaked (Fig. 51) but they enlarge, covering all or part of the leaf, and turn brown or dark brown (Fig. 52). the initial spots are clearly visible on the undersides of leaves.

Figs. 51 and 52. Cassava bacterial blight. Leaf with angular leaf spots viewed with transmitted light (left); angular leaf spots coalescing along the veins and necrosis of the lamina (right). (Maraite and Meyer (1975), *PANS* 21(1): 27–37)

In some cultivars a yellow halo is observed round the spots. The necrosed leaves roll and dry up and after remaining attached to the stem for a short while they fall off. Gum is often exuded from young infected stems, petioles (Fig. 53) and leaf spots, and collects as droplets. On leaves these droplets occur mainly on the undersides along veins or veinlets (Fig. 54). On petioles and stems the gum is exuded from cracks. The gum forms a yellowish glistening scab when dried.

Vascular tissue of infected petioles and stems becomes necrotic and appears as brown strings. Leaves supplied by these vascular tissues wilt and young stem tissues rot. The young green stems rot faster than the older greenish-brown ones which maintain a healthy appearance. The rotting of the young stems causes characteristic dieback symptoms.

Root rot may appear when susceptible cultivars with a good water balance are infected. Roots have a dry vascular necrosis.

On young immature plants complete destruction of the aerial portions may be caused. New shoots are usually produced but these rapidly become infected, especially during the rainy season.

All symptoms appear mainly during the rainy season and vary according to the susceptibility of the variety and the period of infection.

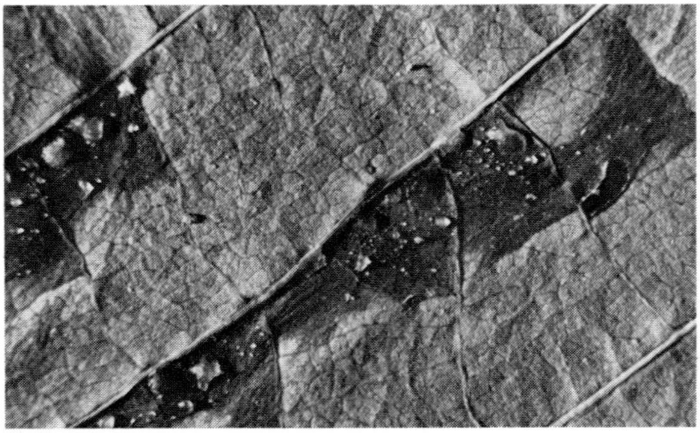

Figs. 53 and 54. Cassava bacterial blight. Droplets of exudate on angular leaf spots on the lower surface of the lamina (above). Droplets of exudate on infected shoot (right). (Maraite and Meyer (1975), *PANS* 21(1): 27–37)

Development and spread

Primary infection is due to the use of infected planting material. Secondary infection results from spread of the bacteria in the gummy exudate by rain splash. Movement of man, animals and insects through the crop and the use of contaminated tools are also likely to spread the disease. In Nigeria the variegated grasshopper (*Zonocerus variegatus*) is considered to be a possible vector. However, rain splash is the most important means of spread over short distances.

Development of the disease is retarded during dry periods and little gum is exuded thus halting the spread of the disease. However, the bacteria remain viable in the plant and are reactivated in later wet periods.

The bacteria enter the plant through stomata or wounds and penetrate to the vascular system from whence they can spread to the rest of the plant. Symptoms generally develop within 7 to 13 days after infection.

Control

Varieties of cassava with field resistance to CBB are available and should be used where possible.

Only healthy planting stock should be used. Clean 'seed' material can be obtained by rooting disease-free stem tips of apparently healthy plants.

Spread of CBB may be delayed by pruning most of the above ground portion of infected plants but the results obtained are variable and complete control would never be obtained by this method. This method is most effective with moderately resistant or resistant cultivars which are only lightly infected.

The CBB organism can only survive in the soil for a month or two enabling land to be cleared of infection by a six month rotation or fallow period before planting disease-free material.

Tools used for cutting planting material should be regularly disinfected to avoid mechanical dissemination of CBB.

Xanthomonas cassavae

Bacterial leaf spot

The symptoms of this disease are angular leaf spots. The causal agent has been considered to be a strain of *Xanthomonas manihotis* but is now considered to be a separate species.

Erwinia cassavae

Bacterial stem rot

The causal agent appears to belong to the *Erwinia carotovora* var. *carotovora* group.

This disease causes shoots to wilt and on woody parts of the stem a characteristic pungent, soft rot or discolouration is observed. Buds on infected stems become necrosed. The bacterium causing the disease enters by wounds, usually caused by the insects *Anastrepha* spp. The bacterium has been isolated from adults and larvae of this insect. The holes caused by *Anastrepha* can be easily located by traces of dried exudate on the stem surface. The best method of control is to use healthy planting material and varieties resistant to the insect vector.

Agrobacterium sp.

Bacterial stem gall

This disease has recently been reported on cassava in Colombia. Infected plants have galls of considerable size on the most woody part of the stem and are stunted. Buds are characteristically present on the galls. The disease is spread by the use of infected planting material.

VIRUS AND MYCOPLASMA-LIKE DISEASES

Several virus diseases of cassava are reported and together they represent a considerable threat to world cassava production. However, there are still many gaps in the available information on these diseases. Details of disease symptoms, losses caused, transmission, host range and causal agents have still to be determined in many cases.

African mosaic disease

Vector *Bemisia* spp.

Distribution

East, West and Central Africa, India; *does not occur in the Americas.*

Symptoms

The disease causes characteristic mosaic symptoms. Young leaves have chlorotic areas and are often deformed (Fig. 55). Leaves may also be reduced in size. When infection is severe leaves are small, misshapen, distorted and have bright yellow areas with normal green tissue in between (Fig. 56). In milder cases of infection leaves may show only a mild chlorotic mosaic. The causal agent has not yet been positively identified but paired isometric particles (30 × 20 nm) have been isolated. A similar mosaic disease has been reported from India. This disease is also transmitted by whitefly but it is not known whether the diseases are the same. It is possible that different strains of mosaic exist but further comparative studies are necessary.

Fig. 55. Cassava African mosaic disease. Leaf showing typical chlorosis and deformation. (E. Terry)

Fig. 56. Cassava African mosaic disease. Chlorosis, reduction of leaf lamina and distortion of severely infected leaf. (Photograph courtesy of Dr R. Williams, IITA, Ibadan, Nigeria. Lozano and Booth (1974), *PANS* 20(1): 30–54)

Control

Estimates of losses caused by African mosaic range from 30–80%. Insect transmission alone does not cause much spread of the disease — man appears to be the main factor.

The use of disease-free planting material and rogueing of infected plants can give excellent control of the disease. Also resistant varieties are available. Areas where this disease does not occur should enforce quarantine measures to ensure that no infected material is imported.

Cassava common mosaic disease

Vector: None recorded

Distribution

Brazil, Colombia, Indonesia, Ivory Coast, Sumatra.

Symptoms

Common mosaic (*CMI/AAB Descriptions of Plant Viruses* No. 90) causes characteristic mosaic symptoms of yellowish areas on the leaf lamina, distortion and stunting of the plant (Fig. 57).

Fig. 57. Cassava common mosaic disease. Leaflets showing mild and severe symptoms. (Lozano and Booth (1974), *PANS* 20(1): 30–54)

The yellowish areas are not as clearly defined as those of African mosaic but otherwise the symptoms of the two diseases are similar. Symptoms of cassava common mosaic may sometimes be confused with those caused by a severe attack of thrips (see page 134). Common mosaic is transmitted mechanically and by grafting but no natural insect vector has been identified. The virus is able to infect other *Manihot* spp., *Euphorbia prunofolia, Chenopodium amaranticolor, C. guinea, Malva parviflora* and *Gossypium hirsutum.*

The causal agent is a virus which has elongated flexuous rods (15 nm in diameter and about 500 nm in length).

Control

The disease has been found to cause losses of 10–20% on the basis of single plant assessments. As there is no insect vector the disease can be easily controlled by the use of clean planting material and rogueing infected plants.

Cassava brown streak disease

Vector: None recorded

Distribution

Cassava brown streak disease occurs only on the east coast of Africa at altitudes under 1,066 m and is uncommon.

Symptoms

Infected plants develop leaf chlorosis and roots become necrotic in storage. Leaf scars persist longer than usual after leaf drop and brown lesions sometimes occur on young stems. Brown streak can also infect other *Manihot* spp. and some other hosts.

As brown streak often occurs in conjunction with mosaic it is difficult to estimate the losses caused but these are thought to be considerable because of the root necrosis.

The virus can be transmitted mechanically and by grafting but no vectors have been identified.

The causal agent has not been positively identified. Elongated rod-shaped particles about 600 nm in length have been associated with the disease. Other workers have isolated paired polyhedral particles (30 x 20 nm). These particles were morphologically similar to those associated with African mosaic and they were related serologically. Further work is needed to clarify etiology of this disease. It has been suggested that the absence of brown streak symptoms except in the East African coast region may be related to varietal tolerance in cassava.

Control

The disease can be controlled effectively, using disease-free planting material and rogueing.

Cassava leaf vein mosaic disease

Vector: None recorded

Distribution

Limited areas of Brazil and Venezuela.

Symptoms

Leaf vein mosaic is of limited economic importance. The characteristic symptoms of the disease are vein-clearing and leaf curling. The disease can be transmitted mechanically and by grafting. Spheroid particles 50-60 nm in diameter have been found associated with the disease. Infected plants should be rogued and only disease-free planting material should be used.

Witches' broom

Vector: None recorded

Distribution

Brazil, Mexico, Peru and Venezuela.

Symptoms

Witches' broom can cause up to 80% reduction in yield. The incidence of infection is low but in affected plantings there is a high percentage of diseased plants.

Diseased plants are stunted with short internodes with excessive proliferation of branches. The disease may cause symptoms of proliferation of shoots from the cutting but in other cases only a few stunted shoots develop. Symptoms may be confused with those caused by heavy thrips infestation in some cases. The disease is transmitted mechanically and by the use of infected cuttings.

Control

Infected plants should be rogued and only disease-free planting material should be used.

Bibliography

BOCK, K. R. and GUTHRIE, E. J. (1976). Recent advances in research on cassava viruses in East Africa. In *Proceedings of the African cassava mosaic workshop.* International Development Research Centre, Ottawa, Canada. Publication 071e. pp. 48.

BRADBURY, J. F. (1975). Bacterial diseases of cassava. *PANS* 21(1): 44.

BRADBURY, J. F. (1977). *CMI Descriptions of Pathogenic Fungi and Bacteria.* No. 559. *Cassava bacterial blight.* Commonwealth Mycological Institute, Kew.

CIAT (1976). *Annual Report for 1975.* Centro Internacional de Agricultura Tropical, Cali, Colombia.

GOLATO, C. (1967). Manioca (*Manihot utilissima* Pohl = *Jatropha manihot.* L.). (Cassava). In *Malattie delle piante coltivate in Somalia.* pp. 20–23. Istituto Agronomico per l'Oltramare, Firenze.

GOLATO, C. and MEOSSI, E. (1966). Una nuova malattia fogliare della manioca in Somalia. *Rivista di Agricoltura Subtropicale e Tropicale* 60: 182–186.

GOLATO, C. and MEOSSI, E. (1971). Una grave infezione fogliare della Manioca in Ghana. *Rivista di Agricoltura Subtropicale e Tropicale* 65 (1–3): 21–26.

KITAJIMA, E. W. and COSTA, A. S. (1964). Elongated particles found associated with cassava brown streak. *East African Agricultural and Forestry Journal* 30: 28–30.

KITAJIMA, E. W. and COSTA, A. S. (1966). Microscopia electronida de tecidos foliaresde mandioca infetados pelo virus do mosaico comum de mandioca. *Bragantia* 25: 23–28.

KITAJIMA, E. W. and COSTA, A. S. (1966). Particulus esferoidais associadas ao virus do mosaico das nervuras da mandioca. *Bragantia* 25(18): 211–221.

KOCH, W. (1974). Tropische Knollenpflanzen und ihre Krankheiten. *Nachrichtenblatt der Deutschen Pflanzenschutzdienstes* 26: 135.

KRAUSZ, J. P. (1976). *The superelongation disease of cassava.* PhD Thesis, Cornell University, Ithaca, New York.

LEU, L. S. (1977). Common bacterial blight of cassava in Taiwan. *Proceedings of the Fourth Symposium of the International Society for Tropical Root Crops,* Cali, Colombia, 1976. pp. 175–179. International Development Research Centre, Ottawa, Canada.

LOZANO, J. C. (1975). Bacterial blight of cassava. *PANS* 21(1): 38–43.

LOZANO, J. C. and BOOTH, R. H. (1973). *The superelongation disease of cassava.* pp. 29. Centro Internacional de Agricultura Tropical, Palmira, Colombia.

LOZANO, J. C. and BOOTH, R. H. (1974). Diseases of cassava (*Manihot esculenta* Crantz). *PANS* 20(1): 30–54.

LOZANO, J. C. and SEQUEIRA, L. (1974). Bacterial blight of cassava in Colombia: etiology. *Phytopathology* 64(1): 74–82.

LOZANO, J. C. and SEQUEIRA, L. (1974). Bacterial blight of cassava in Colombia: epidemiology and control. *Phytopathology* 64(1): 83–88.

LOZANO, J. C. and WHOLEY, D. W. (1974). The production of bacteria-free planting stock of cassava. *World Crops* 26(3): 114–117.

LOZANO, J. C., BELLOTTI, A., SCHOONHOVEN, A. VAN, HOWELER, R., DOLL, J., HOWELL, D. and BATES, T. (1976). *Field problems in cassava.* pp. 127. Centro Internacional de Agricultura Tropical, Cali, Colombia.

MADUEWESI, J. N. C. (1975). Observations on the *Cercospora* leaf-spot disease of cassava (*Manihot esculenta*). *Nigerian Journal of Plant Protection* 1(1): 29—37.

MARAITE, H. and MEYER, J. A. (1975). *Xanthomonas manihotis* (Arthaud-Berthet) Starr, causal agent of bacterial wilt, blight and leaf spots of cassava in Zaire. *PANS* 21(1): 27—37.

NESTEL, B. L. (Ed.) (1976). *African cassava mosaic.* Report of an Interdisciplinary Workshop held at Muguga, Kenya, 19–22 February 1976. pp. 48. International Development Research Centre, Ottawa, Canada.

OLIVEROS, B., LOZANO, J. C. and BOOTH, R. H. (1974). A *Phytophthora* root rot of cassava in Colombia. *Plant Disease Reporter* 58(8): 7035.

RAJAN, K. M., MENON, M. R. and PHILIP, S. (1973). *Phomopsis* blight of tapioca (*Manihot esculenta* Crantz). *Agricultural Research Journal of Kerala* 11(1): 95–96.

SHANMUGAN, N. *et al.* (1964). A die-back disease of tapioca (*Manihot esculenta* Crantz) in Madras, caused by *Glomerella manihotis* Chevaugeon. *South Indian Horticulture* 12(1): 18–23.

TERRY, E. R. (1976). Cassava bacterial blight in Africa. In *The International Exchange and Testing of Cassava Germ Plasm in Africa; Proceedings of an Interdisciplinary Workshop, Ibadan, Nigeria, 1975.* pp. 23–25. International Development Research Centre, Ottawa, Canada.

TERRY, E. R. (1976). Description and evaluation of cassava mosaic disease in Africa. In *The International Exchange and Testing of Cassava Germ Plasm in Africa; Proceedings of an Interdisciplinary Workshop, Ibadan, Nigeria, 1975.* pp.53–54. International Development Research Centre, Ottawa, Canada.

TERRY, E. R. (1977). Factors affecting the incidence of cassava bacterial blight in Africa. *Proceedings of the Fourth Symposium of the International Society for Tropical Root Crops, Cali, Colombia, 1976.* pp. 179–184. International Development Research Centre, Ottawa, Canada.

UNIVERSITY OF GEORGIA (1972). *A literature review and research recommendations on cassava.* University of Georgia, USA.

VYAS, S. C. and JOSHI, L. K. (1974). Leaf blight of cassava (*Manihot utilissima*). *Indian Phytopathology* 27: 422–423.

WILLIAMS, R. J., AGBOOLA, S. D. and SCHNEIDER, R. W. (1973). Bacterial wilt on cassava in Nigeria. *Plant Disease Reporter* 57: 824–827.

WILSON, K. I. and SATHIARAJAN, P. K. (1965). *Phyllosticta* leaf spot of cassava (*Manihot utilissima* Pohl). *Science and Culture* 31(1): 4.

NEMATODES*

Meloidogyne incognita, Meloidogyne spp.

Root-knot nematodes

Distribution

Meloidogyne spp. have been reported on cassava from Brazil, Colombia, Dominican Republic, Hawaii, India, Ivory Coast, Nigeria, Rhodesia, Tanzania, Togo, Trinidad, and USA, and probably occur in all cassava growing regions.

Symptoms

Root-knot nematodes produce the typical swelling or knotting of cassava roots, but there are few reports of them causing serious damage to the crop. In Colombia, cassava roots have been found heavily infected with *Meloidogyne* sp., and, in Trinidad, high populations occur in the roots. However, they seldom caused serious malformations of root systems in Ghana or any reduction in growth, and although galls were numerous they remained small in size. In the Ivory Coast attack by *M. incognita* is considered serious only when cassava was intercropped with good hosts for root-knot, such as eggplant (*Solanum melongena*) and roselle (*Hibiscus sabdariffa*). Two other root-knot species, *M. javanica* and *M. arenaria,* have been found in cassava roots in Nigeria.

Pratylenchus brachyurus, Pratylenchus spp.

Root-lesion nematodes

P. brachyurus is regarded as a possible serious pest of cassava in Togo and has also been recorded in Madagascar and Nigeria. It is a migratory endoparasitic nematode causing damage to the cortical tissues of roots and also predisposing the plant to infection by secondary pathogens; over 40,000 nematodes of the species have been found in 100 g of cassava roots. Other *Pratylenchus* spp. are associated with the crop in various parts of the world.

Other nematodes

Helicotylenchus erythrinae has been frequently encountered as an ectoparasite in the rhizosphere of roots in Togo. *Rotylenchulus reniformis* was found infesting cassava roots in Ghana and the French West Indies, but not causing any reduction in yield. Numerous other plant parasitic species have been reported in or around cassava roots:

*This chapter was written by Dr J. Bridge, ODM Liaison Officer in plant nematology.

Criconemoides spp., *Ditylenchus dipsaci, Helicotylenchus* spp., *Hemicycliophora* spp., *Hoplolaimus pararobustus, Peltamigratus nigeriensis, Scutellonema* spp., *Tetylenchus* sp., *Tylenchus* sp., *Trichodorus* sp., *Xiphinema* spp. There is no evidence that any of these are damaging to cassava.

Control

There are few reports of control measures specifically against nematodes of cassava. Measures taken against nematodes of other root and tuber crops would probably be effective if considered necessary.

In Togo, a 7.2% increase in tuber yield was shown after application of 50 *l*/ha of DBCP to control mainly *Pratylenchus brachyurus.* A number of cassava varieties (Atitogen, Ba Pou II, Agba Tiegba, Agba Boquia, Sodjievi, B25c50) and a pasture legume, *Stylosanthes gracilis,* show some resistance to the species. In Peru, variety Negra Mochera appears most resistant to nematode attack.

Bibliography

BRATHWAITE, C. W. D. (1972). Preliminary studies on plant-parasitic nematodes associated with selected root crops at the University of the West Indies. *Plant Disease Reporter* 56: 1077–1079.

CAVENESS, F. E. (1967). *End of tour progress report on the nematology project.* Ministry of Agriculture and National Resources, Western Region, Nigeria. 137 pp.

EDWARDS, E. E. (1955). Further observations on the occurrence of nematodes of the genus *Meloidogyne* in the Gold Coast. *Journal of Helminthology* 29: 153–170.

GUIRAN DE, G. (1965). Nématodes associés au manioc dans le sud du Togo. *Congress de la Protection des Cultures Tropicales,* Marseille, 1965. pp. 677–680.

HOGGER, C. H. (1971). Plant-parasitic nematodes associated with cassava. *Tropical Root and Tuber Crops Newsletter* 4: 4–9.

LUC, M. (1968). Nematological problems in the former French African tropical territories and Madagascar. In *Tropical nematology.* Edited by G. C. Smart and V. G. Perry. pp. 93–112. University of Florida Press, Gainesville.

INSECT AND MITE PESTS

Cassava can suffer serious insect and mite attacks particularly in the dry season, losing many leaves and shoots, but the plants may recover without the application of pesticides. The use of healthy planting material and resistant cultivars can prevent yield loss from some pests and rain will wash off several others and encourage quick regrowth and recovery. Insecticides should only be applied if the plants are not going to be able to recover from insect attack without insecticide application.

Zonocerus variegatus (Orthoptera: Pyrgomorphidae)

Variegated grasshopper

Distribution

Africa between 15°N and 15°S with the exception of Zambia (CIE Distribution Map 322).

Symptoms

Groups of adult and nymphal grasshoppers defoliate and sometimes strip the bark of the cassava (Figs. 58–60). In Nigeria this occurs during the period November–March. Repeated defoliation reduces yield.

Fig. 58. Cassava being attacked by large numbers of *Zonocerus variegatus*. (COPR)

Fig. 59. Sixth instar *Zonocerus variegatus.* (COPR)

Fig. 60. Adult *Zonocerus variegatus* on cassava. (COPR)

Life cycle

In Nigeria there are two populations of *Z. variegatus;* wet and dry season. The economically important group is the dry season population and the life cycle described here refers to that population. Egg pods, each containing 30–50 eggs, are laid in the soil in shady areas during April. The females collect in large numbers at egg-laying sites and can be seen roosting in the area. Thousands of egg pods may be laid at one site; eggs spend up to 90 days in diapause before embryonic development begins. After 240 days most of the nymphs have hatched. They are basically black in colour with yellow markings. Nymphs pass through 5–7 instars before reaching maturity. Instars 1–3 do not feed on cassava although they may roost upon the plants; they are frequently found feeding on *Eupatorium* spp. Throughout the juvenile instars the insect is very gregarious. By late January the nymphs are reaching adulthood. The adult has drab green elytra, blackish wings and the body is generally green with black, yellow and reddish spots and markings. The wings may be short or reach the seventh abdominal segment.

Control

Cultural

As the egg-laying sites are easily identified when the females gather to lay they should be marked and the eggs dug up during the following 200 days. The eggs will then dry up in the sunshine. Egg-laying sites are unlikely to be more than 5 m^2 and should be dug to a depth of 15 cm. This method of control will only be effective if many farmers in the area participate.

Chemical

BHC and fenitrothion will kill *Z. variegatus* when applied at 0.5% a.i. in water, more than one application may be necessary because of reinvasion. The most satisfactory results will be obtained if first instar nymphs are treated.

***Aleurotrachelus* sp. (Hemiptera: Aleurodidae)**
Aleurothrixus aepim
***Bemisia* spp.**
***Trialeurodes* spp.**

Whiteflies

Distribution

Aleurotrachelus sp.: S. America.

Aleurothrixus aepim: Argentina.

Bemisia hancocki: Cameroon, Chad, Congo (Brazzaville), Israel, Italy, Ivory Coast, Kenya, Malawi, Niger, Nigeria, Rhodesia, Sicily, Sierra Leone, South Africa, Spain, Sudan, Uganda and Zaire.

B. tabaci: Recorded in most parts of the tropics and subtropics. Vector of African mosaic disease.

Trialeurodes manihoti: Brazil.

T. rara: Annobon Is., Cameroon, Central African Republic, Chad, Congo (Brazzaville), Ivory Coast, Nigeria, Sierra Leone, Sudan, Thailand and Uganda.

Symptoms

Severe mottling and curling develop on younger leaves, the older leaves also develop lighter patches of colour and may turn yellow and die. On the underside of the leaves black pupae resembling scale insects covered in wax can be seen. When the plant is shaken adult whiteflies, which are small delicate insects with a white powdery wax deposit on their wings, fly off. Sooty mould may be present if the infestation is heavy.

Life cycle

The stalked eggs are laid on the leaves, usually in semi-circular groups. There are three immature instars, these are flat and oval in shape and they secrete quantities of honeydew. As they develop the legs and antennae degenerate. The fourth instar is the pupal stage, which is usually thicker and more opaque than the previous stages and often covered in waxy deposits. During the early period of the pupal stage the developing adult may be seen inside it. There are about ten generations a year (*B. tabaci*) in Africa.

Control

Cultural

In Africa and Asia where African mosaic disease is a problem virus-free planting material should be used and resistant cultivars chosen if possible. At the Central Tuber Crops Research Institute (Trivandrum) virus infection has been reduced to 1% by this means. In South America cultivars with a moderate resistance to whitefly damage have been identified.

Chemical

Whitefly feeding does not usually cause serious damage, if it is thought necessary to use an insecticide demeton, dicrotophos, dimethoate, methamidophos, omethoate or phosphamidon are recommended for use at the manufacturers recommended rate.

Aonidomytilus albus **(Hemiptera: Diaspididae)**

Cassava stem mussel scale

Distribution

West and East Africa, Argentina, Brazil, India, Madagascar, Mexico, Taiwan, West Indies and USA (Florida) (CIE Distribution Map 81).

Symptoms

Grey, elongate adult scales 2.5–3 mm long and pale nymphs, infest the lower stems of the plant. Infestation may become severe in cassava cuttings kept for planting, the consequent drying up reduces 'germination' success.

Life cycle

About an hour after hatching the nymphs begin to feed, during the next two weeks they secrete a cottony wax cover beneath which moulting and then secretion of a new scale takes place. After another two weeks a second moult occurs. Males subsequently enter the pupal phase and females mature. Male pupation takes less than one week to complete. The entire life cycle takes 5–6 weeks.

Control

Cultural

Clean healthy cuttings should be used.

Chemical

Spraying or dipping cuttings with 0.1% demeton-*S*-methyl, 0.05% parathion or 0.1% malathion is recommended.

Pseudaulacaspis pentagona (Hemiptera: Diaspididae)

White peach scale

Distribution

Cosmopolitan (CIE Distribution Map 58).

Symptoms

Circular papery female scales (2 mm diameter) and long (0.75 mm) male scales can be seen on the stem and leaf bases. Weak or old plants are particularly susceptible.

Life cycle

The nymphs emerge from eggs beneath the female scale and begin to feed. They secrete cottony threads but these disappear when a thin skin is secreted over the nymph. At the first moult the insects become disc-like. Cast skins at each moult are added to the scale. Males moult three times and then pupate. Females mate after the second moult and begin egg laying.

Control

Chemical

Oil emulsion or 0.05% parathion will control this scale.

Parasaissetia nigra (Hemiptera: Coccidae)

Black scale

Distribution

Most warm countries.

Symptoms

Colonies of dark brown to black domed soft scales 2–4 mm long can be seen on the stems (Fig. 61). *Saissetia coffeae, Saissetia miranda* and *Ceroplastes cirripediformis* have also been reported on cassava stems and are similar in appearance to black scale, i.e. hemispherical and waxy in appearance. The scale of *Ceroplastes* is distinctly plated and has six lateral plates and nine seven-sided dorsal plates.

Fig. 61. *Parasaissetia nigra,* the black scale. (G. Swaine (1971) *Agricultural Zoology in Fiji*)

Life cycle

The 'scale' is the female insect, beneath the scale she lays eggs which hatch into crawlers that migrate over the plant before settling down to feed. Although the females do not move around they retain their legs. Female crawlers moult twice before becoming adult and males, which have two wings, moult three times.

Control

Chemical

Cassava cuttings must not be allowed to develop heavy infestations of scale insects since they will produce poor plants when planted or not 'germinate'. Healthy plants will not suffer heavy yield loss due to scale insects. DDT, kerosene emulsion and parathion have been recommended for soft scale control.

Mealybugs (Hemiptera: Pseudococcidae)

Symptoms

Leaves and shoots become distorted and yellow due to mealybugs feeding on them. Adult female mealybugs are segmented soft-bodied insects without wings, about 2.5 mm long and covered in white, waxy meal.

Life cycle

Several species of mealybug are recorded attacking cassava, among them are; *Phenacoccus gossypii, Phenacoccus manihoti* and *Ferrisia virgata.* Generally the female lays groups of 200–400 eggs, which are protected by waxy filaments. In a few hours nymphs, very like the female in appearance but much smaller, emerge. The nymphs pass through several immature stages, if males are present, then they pupate, emerge and mate with the females. The life cycle, egg to adult, may be completed within six weeks.

Control

Cultural

Clean cuttings should always be used.

Chemical

Infected cuttings should be dipped or dusted; aldrin, BHC or DDT have been recommended. BHC, DDT, mephosfolan and parathion sprays or dusts may be applied to the growing plant if it is seriously affected by mealybugs (unless the shoots are to be eaten as a vegetable) and should control the pest.

Vatiga manihotae (Hemiptera: Tingidae)

Cassava lacebug

Distribution

South America.

Symptoms

Early defoliation is caused by large numbers of lacebug adults and nymphs feeding on the underside of the leaves. The leaves develop yellow spots which later turn brown. Lacebugs are recognised by the lacy, reticulate pattern on their pronotum and hemelytra, the prothorax is expanded.

Life cycle

Eggs are laid on the middle and lower leaves and hatch after 8 days. Nymphal instars last 3, 2.5, 3, 3 and 5 days approximately and the adults live about 50 days. Adults are grey and about 3 mm long, the nymphs are paler and smaller.

Control

Chemical

Fenitrothion, applied according to the manufacturers recommendations for lacebugs should control this insect. Losses due to this pest have not been assessed and it may not be economic to spray for lacebug control.

Corynothrips stenopterus (Thysanoptera: Thripidae)
Retithrips syriacus
Scirtothrips manihoti

Thrips

Distribution

C. stenopterus: South America, Barbados, St Vincent and Trinidad.

R. syriacus: Eastern Africa north to Syria, West Africa, Angola, Australia (Northern Territories), Indian sub-continent to the Himalayas and Sri Lanka.

S. manihoti: South America and Trinidad.

Symptoms

Leaves, stems and shoots are deformed and scarred by numerous thrips feeding on them, yellow and brown spots developing on the green tissue. The main growing point sometimes dies, the internodes may be reduced in length and older leaves shrivel and are shed. The dry season is the most common time for severe attacks by thrips but plants may recover in the rainy season. These insects are about 1 mm long, the nymphs are wingless and the adults winged, the narrow wings having very reduced venation and long marginal setae. Adults may be a dark colour, brownish-black; the nymphs are often yellow or red.

Control

Planting resistant cultivars is the best method of control.

Chemical

Dimethoate (160 ml a.i./ha) and thiometon (113 ml a.i/ha) are recommended.

Erinnyis ello (Lepidoptera: Sphingidae)

Cassava hornworm

Distribution

North, central and south America, Antigua, Barbados, British Virgin Islands, Dominica, Dominican Republic, Grenada, Guadeloupe, Jamaica, Martinique, Montserrat, St Kitts, St Lucia, St Vincent, Trinidad and Tobago.

Symptoms

Defoliation is caused by hornworms that are similar in appearance to the taro hornworm (Fig. 94). If attacked by a heavy infestation during the first two months of growth some plants may die and the crop yield will suffer. Several authors have noted that tuber quality is lost, the starch content is reduced and the tubers are watery, thus increasing the processing costs. On fertile soil even 20% defoliation after the second month of growth has very little effect on yield. The effect would be more severe under poor soil conditions.

Life cycle

Green spherical eggs 1.5 mm in diameter are laid, singly or in groups on both surfaces of the leaves or on the petioles and hatch within a week. At first the larvae feed on very young leaves and shoots, as they develop they eat older leaves but not the very old basal leaves as these are toxic to them. There are five larval instars. The fifth is 80–90 mm long and may be green to brownish-black in colour. Their colour is determined by larval crowding, food and some unknown interlarval stimuli. After 2 weeks the larvae pupate for 2–3 weeks in leaf litter or soil. The pupae are dark brown and about 45 mm long.

Adult *E. ello* have grey/black banding on the abdomen, grey (female) and grey-brown with a black stripe (male) forewings and grey hindwings. They have a wingspan close to 10 cm. There are 2–8 generations per year.

During the day adults rest on plants. The genus *Erinnyis* is apparently restricted to host plants that secrete latex. *Erinnyis ello* has been found breeding on cassava, various

Euphorbia spp. and *Cnidoscolus* spp.; *Hevea brasiliensis, Carica papaya,* some Solanaceae, *Lucuma caimito, Allamanda cathartica, Aranjia sericifera, Allium cepa, Yucca* sp., *Gossypium herbaceum, Ficus* sp. and *Psidium* sp.

Control

Biological

In Colombia *Trichogramma fasciatum,* an egg parasite, kills 60–100% of *E. ello* eggs on cassava and with a predatory wasp *Polistes* sp. usually achieves excellent control. *Polistes canadensis* predation has been increased at CIAT by placing two 2 m high 'wasp houses' covered in palm leaves in the cassava field (Fig. 62). Initially each house contained five wasp nests and these are increased to about 40 after six months.

Numerous other parasites and predators have been recorded attacking *E. ello* in various countries. If the crop is not treated with an insecticide that kills the beneficial insects then good biological control may be expected. Hand collection of larvae is useful in small areas of the crop. Construction of pits full of leaf litter as pupating sites to attract the mature larvae and subsequent collection and destruction of the pupae has also been recommended. Latex-secreting weeds should be destroyed.

If possible, *Bacillus thuringiensis* sprays should be used to reduce *E. ello* larval numbers without damaging the predator and parasite population. Treatment of the first three instars is the most effective.

Fig. 62. *Polistes* nest in a wasp house. (CIAT)

Chemical

Soap solutions, lead arsenate (400 g/100 *l* water plus 100 ml light oil). Paris green (500 g/500 *l* water plus 3 kg lime as a sticker) or trichlorphon 2.5% dust are alternatives recommended if *Bacillus thuringiensis* is not available.

Anastrepha manihoti (Diptera: Tephritidae)
Anastrepha pickeli
Fruitflies

Distribution

Anastrepha manihoti: Brazil, Panama and Peru.

Anastrepha pickeli: Argentina, Brazil, Panama and Peru.

Symptoms

Stems 'bleed' a white exudate due to yellowish-white legless larvae boring in the pith 10–20 cm below the growing point. Infection with bacterial stem rot (see p. 117) is sometimes associated with fruitfly attack. Heavy attacks and the combined effect of insect and pathogen can lead to the death of the main growing point, causing branching and reduced yield.

Life cycle

The flies are yellowish-brown with distinct dark markings on their wings (Fig. 63). When at rest they commonly extend the wings and raise and lower them. They lay the eggs in the stem. The larvae are about 5 mm long, wider at the posterior end than at the head, where the sclerotised mouthparts may be seen. Pupation takes place in the soil.

Control

Chemical

Chemical control is probably only necessary when young plants (2–5 months) are severely attacked. The larvae can be killed in the stem by spraying with fenthion at the manufacturer's recommended rates.

Fig. 63. Adult *Anastrepha* sp., a fruitfly, on cassava. (CIAT)

Silba pendula (Diptera: Lonchaeidae)
Carpolonchaea chalybea

Shootflies

Distribution

South America.

Symptoms

The growing point becomes brown and deformed due to white larvae burrowing inside it and there is frequently a whitish-yellow exudate from the damaged shoot. The growing point may die and secondary shoots develop, which may also be attacked. Attacks take place at the beginning of the rainy season.

Life cycle

The adult is a shiny blue fly. Eggs are laid among the folded leaves of the growing plant and after 4 days the larvae emerge and bore into the plant to feed. The larval stage lasts 20–23 days and pupation, which takes place in the soil, lasts 25 days.

Control

Cultural

Planting when the shootfly population is low and the use of resistant cultivars are recommended. In Brazil the varieties 1418IAC and Ouro do Vale were found to be highly resistant to this pest.

Chemical

Trials in which the growing point was removed to simulate shootfly damage have indicated that older plants do not suffer economic losses from shootfly, therefore pesticides should only be used to protect young plants. The larvae are reported to be difficult to control and some attempts have been made to bait the adults with mixtures of DDT 0.25% or malathion 0.15% and molasses (5%). Diazinon, dicrotophos, dimethoate, methamidophos or monocrotophos have been recommended, at the manufacturer's recommended rates, for larval control.

Atta **sp.** **(Hymenoptera: Formicidae)**
Acromyrmex **sp.**

Leaf-cutting ants

Distribution

Caribbean, Central and South America.

Symptoms

Semi-circular pieces of leaves are cut out and carried away by worker ants. The ants' nests can be identified by the piles of dirt around the entrance holes. Whole plots of cassava can be defoliated and even buds may be taken.

Life cycle

These ants live in nests up to 10 m in diameter extending several metres below ground level. A nest is made up of brood chambers and chambers in which fungus gardens are maintained. The leaves are chewed and used as 'soil' for the fungus on which the ants feed. The ant colony consists of a queen, various castes and the brood.

Control

Chemical

Poisoned baits applied over a wide area during the dry season are the most satisfactory method of control. The foraging ants carry the poison back to the nest where the rest of the colony is affected and dies. The bait must be attractive and remain so for 2–4 weeks and the poison must be slow acting, to give the foragers time to carry it back to the nest. If it is necessary to apply bait in the wet season then it should be waterproofed to prevent decay, 2% methyl-trichloro-silane is recommended for this purpose. Baits consisting of 0.4% insecticide + 8.5% soyabean oil + 91.1% dried citrus meal are recommended, mirex or aldrin are the preferred insecticides. The insecticide should be dissolved in the soyabean oil and then well mixed with the citrus meal. Commercially prepared baits are available. The farmer should protect his hands while applying bait.

Fumigating the nests with carbon disulphide, sulphur smoke or arsenates will also destroy the colony or alternatively the nest entrance may be treated with aldrin, BHC or dieldrin in solution.

Mononychellus tanajoa (Acarina: Tetranychidae)

Tanajoa, green cassava mite, green spider mite

Distribution

Brazil, Burundi, Congo (Brazzaville), Colombia, Guyana, Kenya, Paraguay, southern Sudan, Tanzania, Trinidad, Uganda and Zaire.

Symptoms

The growing point, young leaves and stem are attacked by green mites. Leaves develop mosaic symptoms and yellow spots, darken, dry up and become deformed (Fig. 64). Shoots and leaves die progressively from the top, the plants become stunted and may branch. *Mononychellus mcgregori* causes similar damage. Damage is most severe during the dry season.

Fig. 64. Typical damage caused by *Mononychellus* sp. on the growing point of the cassava plant. (CIAT (1975) *Cassava production systems*)

Life cycle

In the field eggs are laid singly on the lower surface of leaves. In the laboratory egg, larval, protonymphal and deutonymphal stages last 3–4, 1–2, 1–2, and 2–3 days respectively. During heavy rainfall all stages of the mite may be washed off the plant. Females disperse in the wind after lowering themselves on silken threads. This mite does not construct webbing.

Control

Biological

Predaceous mites and staphylinid beetles in Central America and the West Indies feed on green cassava mite.

Chemical

See Table 13.

TABLE 13. RECOMMENDATIONS FOR CHEMICAL CONTROL OF MITES

Pesticide	Formulation	Active ingredient	Application
Binapacryl	e.c. or w.p.	0.04%	Apply as soon as eggs start hatching and repeat as necessary
Chlorobenzilate		0.05%	
Diazinon	Follow the manufacturer's recommendations		
Dicofol	Follow the manufacturer's recommendations		
Dimethoate		0.03%	
Lime sulphur		0.01%	
Mancozeb		0.16%	
Methamidophos	Follow the manufacturer's recommendations		
Naled	e.c.	0.001%	
Parathion		0.03%	
Phosalone		0.07%	

***Tetranychus cinnabarinus* (Acarina: Tetranychidae)**

Spider mite, carmine spider mite

Distribution

Throughout the semitropics and in protected cultivation in temperate zones.

Symptoms

Lower leaves develop yellow dots along the veins, these dots become reddish and if the dry season is prolonged the infestation spreads to higher leaves. Leaves dry up and fall and some plants die. The mites responsible are brownish-red with dark lateral internal markings, they produce webs (Fig. 65).

Tetranychus marianae, widespread in the Pacific Islands, produces similar symptoms.

Fig. 65. Webbing and damage caused by a species of *Tetranychus* under controlled environmental conditions. (CIAT (1975) *Cassava production systems*)

Life cycle

Reproduction takes place during dry weather, the optimum temperature for development is 32°C. Single eggs are laid on the under surface of the leaves or on the webbing. Development, egg to adult takes about 17 days and there may be 19 broods per year.

Control

Cultural

Mite tolerant cultivars have been identified and should be planted if available.

Chemical

See Table 13.

Oligonychus (Homonychus) peruvianus **(Acarina: Tetranychidae)**

Distribution

Guatemala, Mexico, Peru, Trinidad, USA (California and Texas) and Venezuala.

Symptoms

Older leaves are attacked and whitish spots (webbing) develop on the underside close to veins, on the upper surface these spots are yellow-brown. The mites feed under the webbing.

Life cycle

Egg laying and nymphal development take place beneath the webbing.

Control

Chemical

See Table 13.

Eutetranychus orientalis and *Oligonychus* (*Pritchardinychus*) *biharensis* are reported from India, feeding on the dorsal leaf surface during the period December–April and causing losses in yield as high as 20%. In Nigeria the red spider mite *Oligonychus* (*Reckiella*) *gossypii* is reported on cassava.

Minor pests

Gallmidges are reported attacking cassava in several countries. Adult female cecidomyids lay their eggs on the leaves and the emerging larvae begin to feed, their attack stimulates abnormal cell growth on the leaf and red galls develop on the upper surface. Affected leaves should be removed and destroyed once a week to prevent the pest becoming a serious problem. The larvae of various beetles, for example *Leucopholis coneophora* (in India) attack the cuttings and young roots. These larvae, or white grubs, usually Scarabaeidae or Cerambycidae can be seen when preparing the land for planting. Where they are present it may be necessary to treat the soil beneath the cutting with aldrin or carbofuran. In India *L. coneophora* adults are killed by laying branches of leafy trees about the field and spraying them with BHC, endosulfan or quinalphos, as the adults emerge after the first monsoon shower (April, May or June) they go to feed on the leaves and die. When the plants wilt due to these larvae feeding the soil can be drenched with 0.03–0.05% BHC.

Cutworms, lepidopteran larvae, may attack the stems above and below the soil surface, they can girdle the stem and destroy the roots, treatment of the soil with aldrin or carbofuran can be carried out or poison baits consisting of 10 kg sawdust + 8–10 *l* water + 500 g sugar or molasses + 100 g trichlorphon may be scattered over 0.5–1.0 ha. The weevil *Coelosternus* and the pyralid *Chilomina clarkei* are reported as stem borers, their presence can be detected by the excreta and sawdust which drops from their tunnels. Cutting off the infected branches is the recommended control method.

Bibliography

ALBUQUERQUE, M. D. (1977). Mealy bug attack on cassava in Amazonia. *Proceedings of the fourth Symposium of the International Society for Tropical Root Crops,* Cali, Colombia, 1976. p. 207. International Development Research Centre, Ottawa, Canada.

ANON. (1964). A mandarova, a pior praga da mandioca; aspectos biologicos, combate con dipterex pó 2.5% ou Folidol EM. 7.5% + 30% DDT. *Correio Agro-Pecuario* (Brazil) 4(2): 18–19.

ANON. (1975). Sistema de producão da mandioca reconcavo e sul da Bahia *Serie extensão* – ano 1, No. 1 Cruz das almas – Bahia, Brazil. pp. 30.

BELLOTTI, A. and SCHOONHOVEN, A. VAN (1977). World distribution, identification, and control of cassava pests. In *Proceedings of the Fourth Symposium of the International Society for Tropical Root Crops,* Cali, Colombia, 1976. pp. 188–193. International Development Research Centre, Ottawa, Canada.

BELLOTTI, A., SCHOONHOVEN, A. VAN and PENA, J. (1975). Insectos que atacan la yuca. In *Curso sobre producción de yuca.* Instituto Colombiano Agropecuario, Medellin, Regional 4, pp. 45–60.

BRETONI, A. DE W. (1924). Sobre el mandarová de la mandioca (*Erinnyis ello* L.) *Revista de la Sociedad Cientifica del Paraguay* 1(6): 91.

CIAT (1975). *Cassava production systems.* pp. 57. Centro Internacional de Agricultura Tropical, Cali, Colombia.

CIAT (1975). *Annual Report for 1974.* Centro Internacional de Agricultura Tropical, Cali, Colombia.

CIAT (1976). *Cassava production systems program.* pp. 76. Centro Internacional de Agricultura Tropical, Cali, Colombia.

CORSEUIL, E. (1954). Mandarová da mandioca. *Boletim do Campo* 10(75): 3–8.

FENNAH, R. G. (1947). *The insect pests of food crops in the Lesser Antilles.* pp. 24. British West Indies Department of Agriculture.

FONSECA, J. P. DA. (1943). *Mandarová da mandioca.* pp. 14. Sao Paulo Secretaria de Agricultura, Instituto Biologico. Folheto no. 98.

FONSECA, J. P. DA. (1943). Combate ao mandarová da mandioca. *Notas Agricolas* (*Brazil*) 6: 389–394.

LEUSCHNER, K. (1976). Major pests of cassava in Africa and preliminary guidelines for screening of resistance. In *The International Exchange and Testing of Cassava Germ Plasm in Africa; Proceedings of an Interdisciplinary Workshop,* Ibadan, Nigeria, 1975. pp. 55–56. International Development Research Centre, Ottawa, Canada.

LEWIS, T. (1972). Aerial baiting to control leaf-cutting ants. *PANS* 18(1): 71–74.

LEWIS, T. (1973). *Thrips their biology, ecology and economic importance.* pp. 349. Academic Press Inc., (London).

LEWIS, T. and PHILLIPS, F. T. (1973). Aerial baiting to control leaf-cutting ants (Formicidae, Attini) in Trinidad. I. The bait, its production, and the effects of weathering on attractiveness and persistence of toxicants. *Bulletin of Entomological Research.* 63: 263–273.

LOZANO, J. C., BELLOTTI, A., SCHOONHOVEN, A. VAN, HOWELER, R., DOLL, J., HOWELL, D. and BATES, T. (1976). *Field problems in cassava.* pp. 127. Centro Internacional de Agricultura Tropical, Cali, Colombia.

LYON, W. F. (1973). A plant feeding mite *Mononychellus tanajoa* (Bondar) (Acarina: Tetranychidae) new to the African continent threatens cassava (*Manihot esculenta* Crantz) in Uganda, East Africa. *PANS* 19(1): 36–37.

LYON, W. F. (1974). A green cassava mite recently found in Africa. *Plant Protection Bulletin* 22(1): 11–13.

MINISTRY OF OVERSEAS DEVELOPMENT/UNIVERSITY OF IBADAN. (1974). *Control of* Zonocerus variegatus *(L.) in Nigeria. Interim report January 1972–July 1973.* pp. 17. Centre for Overseas Pest Research, London.

MINISTRY OF OVERSEAS DEVELOPMENT/UNIVERSITY OF IBADAN. (1975). *Control of* Zonocerus variegatus *(L.) in Nigeria. Second interim report* 1973–74. pp. 15. Centre for Overseas Pest Research, London.

MINISTRY OF OVERSEAS DEVELOPMENT/UNIVERSITY OF IBADAN. (1977). *Control of* Zonocerus variegatus *(L.) in Nigeria Final report and recommendations.* pp. 15. Centre for Overseas Pest Research, London.

MATTOS, P. L. P. DE, GOMES, J. DE C. and MATOS, A. P. LE. (1973). *Cultura da Manioca. Circular No.* 27. pp. 13. Instituto de Pesquisas Agropecuarias do Leste.

MOUND, L. A., and HALSEY, S. H. (1978). *Whiteflies of the world.* British Museum (Natural History), London.

NORMANHA, E. S. (1946). *O. mandarová da mandioca.* p. 3. Instituto Agronômico, Campinas, Brasil.

NORMANHA, E. S. (1965). Comefolha, prejudica raiz. *Coopercotia* 22(190): 39–40.

PHILLIPS, F. T. (1974). The weathering of ant baits containing aldrin, dieldrin and mirex in Trinidad. *Pesticide Science* 5(2): 147–157.

PHILLIPS, F. T. and LEWIS, T. (1973). Current trends in the development of baits against leaf-cutting ants. *PANS* 19(4): 483–487.

PILLAI, K. S. (1975). Pests of tapioca and their control. In *Short specialized training course on tapioca production technology,* 1975. Central Tuber Crops Research Institute, Sreekaryam, Trivandrum.

RAO, Y. R. V. J. and PILLAI, K. S. (1972). Effect of some insecticides on the scale insect, *Aonidomytilus albus* Ckll., infesting tapioca stems in storage. *Indian Journal of Entomology* 34(2): 181–182.

ROSSETO, C. (1970). principais pragas da mandioca no Estado de Sao Paulo, *Encontro do Engenheiros Agronômos Pesquisadores em Mandioca dos Paises Andinos e do Estado de Sao Paulo, Campinas, Brasil.* pp. 90–95.

SCHOONHOVEN, A. VAN. (1974). Resistance to thrips damage in cassava. *Journal of economic Entomology* 67: 728–730.

SCHOONHOVEN, A. VAN and PENA, J. E. (1976). Estimation of yield losses in cassava following attack from thrips. *Journal of economic Entomology* 69(4): 514–516.

SIVAGAMI, R. and NAGARAJA-RAO, K. R. (1967). Control of the tapioca scale *Aonidomytilus albus* Ckll. *Madras Agricultural Journal* 54(6): 325–327.

VILLAMIZER, M. R. (1974). *Ciclo biologico y morfologia de* Vatiga (Leptopharsa) manihotae (*Drake*) (*Hemiptera: Tingidae*), *chinche de encaje en yuca.* Tesis Ing. Agr. Universidad Nacional de Colombia, Facultad de Ciencias Agropecuarias, Palmira. pp. 40.

WILSON, T. H. (1975). A monograph of the sub family Panchaetothripinae (Thysanoptera: Thripidae). *Memoirs of the American Entomological Institute* No. 23.

WINDER, J. A. (1976). Ecology and control of *Erinnyis ello* and *E. alope,* important insect pests in the New World. *PANS* 22(4): 449–466.

YÉPEZ, F. F. and TERÁN, B. J. B. (1973). Presencia de *Chiloma clarkei* (Amsel) y *Chilozela bifilalis* (Hampson), (Lepidoptera, Pyralidae) en yuca (*Manihot esculenta* Crantz) en Venezuela. *Agronomia Tropical* 23(4): 407–411.

YAMS

DISEASES

Introduction

The information on yam pathology available in the literature is sparse when compared with that on other more economically important crops. This chapter may well reflect this lack of information and hopefully encourage further interest. Numerous pathogens have been recorded on yams but none is noted as causing generally significant pre-harvest losses although damage may be serious locally. The distributions may in fact be wider than quoted.

FUNGAL AND BACTERIAL DISEASES

Glomerella cingulata

Anthracnose

Distribution

Caribbean, East Africa, Far East, India, West Africa.

Symptoms

Symptoms of the disease vary between species of yams. In *Dioscorea alata* the first signs of infection are small brown spots (2–5 mm in diameter), sometimes with a yellow edge, which appear mainly on the lower leaves and stems. The spots enlarge and spread giving the stem a blackened appearance and causing leaves to wither. Brown to black macroscopic acervuli (fruiting bodies) may be observed on the lesions of the leaves and stem, particularly during the rains, giving the appearance of dark concentric rings on the lesions. Sometimes secondary infection occurs and the leaf is destroyed.

Stems are also attacked, but less frequently. If the infection reaches the stele death of the plant can result. Attacks on the stem generally occur as a result of leaf infections but they can occur as a result of direct penetration near the terminal bud which leads to rapid death of the plant. Direct infection can also occur at ground level.

D. bulbifera is particularly susceptible to anthracnose. The symptoms are similar to those on *D. alata.* However a greater number of leaves are attacked at the same time and necrosis is more rapid. On the stem there are as many infected zones around the points of attachment of the petioles as there are infected leaves. These stems dry up and the plants die.

On *D. esculenta* the symptoms are as obvious on the stems as on the leaves. Leaf infection first appears at the apex and edge of the leaf in the region with the highest density of stomata. The brown spots bordered with yellow, which appear first, enlarge and develop concentric circles of light and dark brown. The infection spreads across the leaf and reaches the petiole. When the petiole becomes infected the leaf droops and then falls off. The stem can become infected via the petiole. In very humid conditions direct infection of the stem apices can occur. The first symptoms appear 20 cm from the top and extend upwards to the terminal bud, which dies, and downwards where the limit of the infection is marked by a reddish brown line. The affected stem rapidly loses its chlorophyll, fades, dries and shrivels and then dies. The older parts of the stem do not become infected. Anthracnose is generally confined to the upper 70 cm.

The reduction in yield caused by anthracnose can be particularly serious in the more humid regions. *Dioscorea alata* is very susceptible but D. esculenta is more resistant. *D. cayenensis* and *D. rotundata* are only slightly susceptible, even in humid forest regions, their leaves only develop a few brown spots (similar to those described for *D. alata*) and stems are not attacked.

Development and spread

Anthracnose overseasons on crop debris as acervuli. Infection is favoured by wet and humid conditions. Many acervuli of the *Gloeosporium* type which become erumpent later, are produced on lesions. Acervuli remain devoid of setae when humidity is over 95% but at lower humidities sterile setae are mixed with conidiophores converting the acervuli to the *Colletotrichum* type. In very dry conditions only setae are formed. Spores are mainly dispersed by rain splash and the disease is therefore most severe when there are heavy rains.

Control

Crop debris should be removed as it is a major source of inoculum. Species or varieties of yam which are less susceptible should be planted — especially in the humid regions where the disease is common. Various fungicides have been tested for control of anthracnose. In Fiji maneb at 3.32 g a.i./*l* every 10 days was effective. Benomyl alone or applied alternately with mancozeb on a weekly cycle has also been shown to give good control and increase yields. Benomyl + propineb is also effective. (Benomyl should be used with caution as resistant strains of the pathogen may develop.) In one trial zineb, ferbam and ziram sprays were shown to give 50%, 41% and 30% yield increases respectively. Copper fungicides caused some phytotoxicity.

Cercospora brasiliensis, C. carbonacea, C. pachyderma, Cercospora* sp., *Phaeoramularia dioscoreae, Pseudocercospora contraria

Leaf spots

Distribution

World wide on all species of yams.

Symptoms

Several *Cercospora* species and some species which were formerly classified as *Cercospora* are recorded as causal agents of leaf spots on yams. Generally the symptoms are brownish spots which darken with age and coalesce as the disease progresses (Fig. 66).

Fig. 66. Leaf spot with yellow halo, probably caused by species of *Curvularia* or *Cercospora.* (E. Terry)

Phaeoramularia dioscoreae causes small brown spots which sometimes coalesce and has been recorded in India, USA and South America.

C. carbonacea is the most commonly recorded causal agent and causes large blackish-brown to brown spots which may spread over the whole leaf and has been recorded in Barbados, Burma, Cuba, Ghana, Grenada, India, Jamaica, Nigeria, Puerto Rico, Sierra Leone, St. Vincent, Tanzania, Togo, Trinidad and Venezuela.

Cercospora sp. was isolated in the Ivory Coast from leaves with rounded grey spots which are dry in the centre, and surrounded by a persistent reddish margin.

Pseudocercospora contraria occurs in Sierra Leone.

C. pachyderma occurs in Burma, China, India, Japan, Malaysia, Sarawak, Sierra Leone and Uganda.

C. brasiliensis occurs in Brazil and Venezuela.

A leaf spot caused by *Cercospora* sp. is reported from Fiji. Initially the spots are small and brown or yellowish. Later they enlarge to angular black spots about 6 mm across. These may coalesce and affect the whole leaf surface. The centres of the spots sometimes fall out. Spots may also appear on the petioles and stems.

Damage due to these leaf spots is not generally serious.

Development and spread

The fungus overseasons on crop debris. Conidia are produced on both sides of infected leaves and are spread by rain. The disease is more common during warm and wet weather.

Control

Control measures are not necessary unless the disease is severe. Sprays with maneb or captan at 2.07 g/*l* every 10 days have been found to be effective.

Corticium rolfsii

Leaf spot, neck rot

Distribution

Australia, Ivory Coast, Jamaica, Sierra Leone etc. Widespread.

Symptoms

This is one of the most important foliar diseases of yams and can cause serious damage when stems are affected (*CMI Descriptions of Pathogenic Fungi and Bacteria* No. 410). The foliar lesions are very different on the various species of yams.

On *Dioscorea alata* spots are circular, greyish-beige in the centre bordered with light to purplish-brown surrounded by a yellowing halo and can reach 2 cm in diameter. Sometimes the spots run together and often the centres fall out. On the underside of the leaves small reddish-brown sclerotia can be observed in the centre of the spots. If the humidity is high white hyphae begin to grow from the sclerotia and radiate towards the edge. A bright greasy looking border shows the area affected by the fungus. Many sclerotia, which are white at first, then reddish-brown and finally brown, appear on the underside of necrotic leaves. On *D. esculenta, D. cayenensis* and *D. bulbifera* the necrosis is first seen as a greyish-beige or pinkish-beige spot limited by a dark beige ring surrounded by a yellow halo. Then the whole of the leaf takes on a beige or light coffee colour with dark bands. The sclerotia appear early and in great number on the underside of the spots and then, sometimes, on the surface. The petiole can become affected after the leaf.

Leaves of *D. dumetorum* are not attacked. This seems to be associated with their abundant hairs.

Leaves are infected directly and individually without having to be in contact with the soil which harbours *C. rolfsii.* Rain splashing can result in the transfer of enough sclerotia to ensure infection. When the infection starts in the base of the stem the first symptom is browning of the lower leaves. Later the lower part of the stem turns black and the aerial parts fade. When the tuber is dug up total disorganisation of the tissues in the neck (the part where the tuber joins the stem) is observed.

Foliar infections are only observed during the rainy season and are not usually very serious. Development of the disease is halted by the start of the dry season.

Neck infection is more serious and causes death in *D. bulbifera* and *D. alata.* However young stems of *D. alata* possess some resistance as they can produce new tissue which prevents the stem from being girdled. *D. esculenta* is resistant or at least only slightly susceptible. Secondary infection with several other fungi usually follows neck infection by *C. rolfsii.*

Control

The presence of foliar lesions indicates a high soil population of *C. rolfsii* which can attack all the crops grown in association with or in rotation with yams (e.g. groundnuts, cotton, maize, cassava and rice). If foliar *C. rolfsii* infection of yams has been observed it is wise to rotate with resistant cultivars of the other crops. This is particularly important in the case of groundnuts which are severely attacked. Simple cultural practices can avoid or reduce the damage. Training the vines up supports to keep the leaves clear of the soil can appreciably reduce the chance of infection.

Once neck infection has appeared it is too late to apply effective control measures. Chemical control seems inapplicable. Use of the resistant species, *D. esculenta*, instead of the more susceptible species could be encouraged in districts where infection is widespread.

Pestalotiopsis cruenta

Leaf spot

Distribution

Ivory Coast.

Symptoms

Reddish-brown spots appear on the leaves and more rarely on the stems. These spots first appear as reddish-grey pinpoints on the upper leaf surfaces. These points enlarge and coalesce to form spots which are dark brown or have reddish zones. The acervuli are scattered. Symptoms are similar on all yam species.

On young branches *P. cruenta* forms brown spots which enlarge and cause the lower part of the stem to dry up.

This disease mainly affects *D. esculenta* in very bad growing conditions and can be considered a weak parasite. On *D. alata* and *D. cayenensis* it appears as a secondary infection after *Glomerella cingulata* or *Phyllosticta dioscoreae.*

Colletotrichum capsici

Leaf spot

This disease has been recorded on yams in Malaya, Ivory Coast and Formosa (*CMI Descriptions of Pathogenic Fungi and Bacteria* No. 317).

Large grey areas bordered with black and dotted with black acervuli are observed on lower leaves. The disease is of minor importance and no special control measures are recommended.

Thanatephorus cucumeris

Leaf blight

Distribution

This disease is recorded from India.

Fig. 67. Leaf blight, caused by an unknown pathogen. (E. Terry)

Symptoms

The first symptoms of this disease (*CMI Descriptions of Pathogenic Fungi and Bacteria* No. 406) are observed on the lower leaves which are near the soil. The infected areas are clearly defined and have a water-soaked appearance. The infection spreads rapidly over the leaf lamina and the infected areas become flaccid, collapse and turn brown (Fig. 67). A concentric light and dark brown colour zonation is often observed. After collapse of the leaf the infection may spread via the petiole and the stem to other leaves. Mycelial strands are often observed on the surface of infected leaves. This disease is not recorded as being of economic importance and no control measures have been recommended.

Phyllosticta dioscoreae

Leaf spot

Distribution

Formosa, Ghana, Ivory Coast, USA.

Symptoms

Leaves of all ages are attacked but attacks are more frequent on lower leaves. The infection starts on any part of the leaf as isolated, brown, punctiform spots which are visible on both surfaces of the leaf. The centres of the spots become depressed and greyish-brown to grey in colour, while the edges are dark brown. When growing leaves are attacked, holes are formed in the centre of the spots. Pycnidia, which appear on both surfaces of the leaves during the rainy season, are scattered, and globular or spheroid and open more often on the upper surface.

Often there is secondary infection with several other species of fungi. If infection is confined to older leaves no significant yield reduction is caused, and control measures have not been developed.

Phyllosticta dioscoricola

Leaf spot

Distribution

Ivory Coast.

Symptoms

This disease occurs in both forest and savanna regions on *Dioscorea alata, D. bulbifera, D. cayenensis* and *D. esculenta*; *D. cayenensis* being attacked the most severely. *D. bulbifera* and *D. alata* are only slightly affected. The initial symptoms are numerous spots, which are brownish-red to light buff in colour and bordered with dark brownish-red. On leaves of *D. esculenta* the spots are less numerous but larger. There is little host reaction and the border is very narrow with an irregular outline. Symptoms are similar on *D. bulbifera*. Lesions are rare on *D. alata*. Pycnidia are visible on the upper surface of the lamina as black specks.

The development of the disease appears to depend greatly on the stage of development of the plant, the cultural conditions and the climate. Infection reduces the number and volume of tubers of *D. esculenta,* but the magnitude of the loss is not recorded and no control measures are mentioned in the literature.

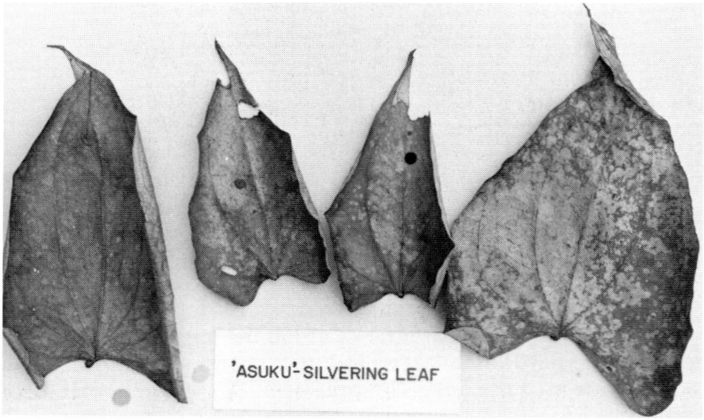

Fig. 68. Silvering leaf symptoms caused by *Botryodiplodia* sp. (E. Terry)

Minor leaf diseases

Phyllosticta grafficina leaf spot has been recorded from Nigeria.

Gloeosporium pestis has been recorded widely as the causal agent of a leaf spot and wilt of leaves and young stems.

Botryodiplodia theobromae (*CMI Descriptions of Pathogenic Fungi and Bacteria* No. 519) is mainly important as a tuber rot but can also cause necrosis of the leaves (Fig. 68).

Curvularia spp. have been associated with leaf spots in many countries (Fig. 69).

Uredo dioscoreae is recorded as the causal agent of yam rust in Kenya.

Phakopsora dioscoreae and *Uromyces combreti* are recorded as causal agents of yam rust in Burma.

Goplana dioscoreae causes a yam rust which has a worldwide distribution.

Agrobacterium tumefaciens (*CMI Descriptions of Pathogenic Fungi and Bacteria* No. 42) causes a crown gall disease and has been recorded in the West Indies but not elsewhere.

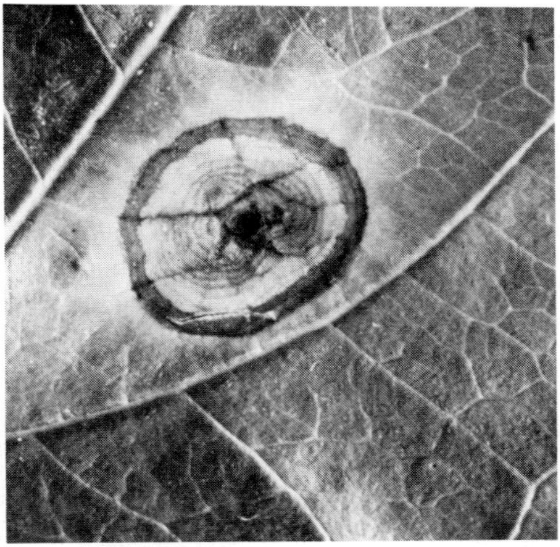

Fig. 69. Zonate leaf spot on *Dioscorea rotundata* caused by *Curvularia* sp. (E. Terry)

Phylleutypa dioscoreae is associated with a witches' broom disease in several yam species. The disease occurs commonly but not widely in West Africa.

Botryodiplodia theobromae

Tuber rot

Distribution

Brazil, Ghana, Ivory Coast and Nigeria.

Symptoms

B. theobromae (*CMI Descriptions of Pathogenic Fungi and Bacteria* No. 519) is one of the most important causal agents of tuber rot and can be associated with wet rot, soft rot and brown dry rot: infection often occurs in conjunction with nematode or insect attack (Figs. 70 and 71). The wet rot results in the tuber tissues disintegrating into a watery mash, while tissues are slightly softened and pink in colour with the soft rot. With the dry rot the tissue becomes brown, dries and finally crumbles away as a brown powder. The symptom differences are probably associated with variations in the species of secondary pathogens.

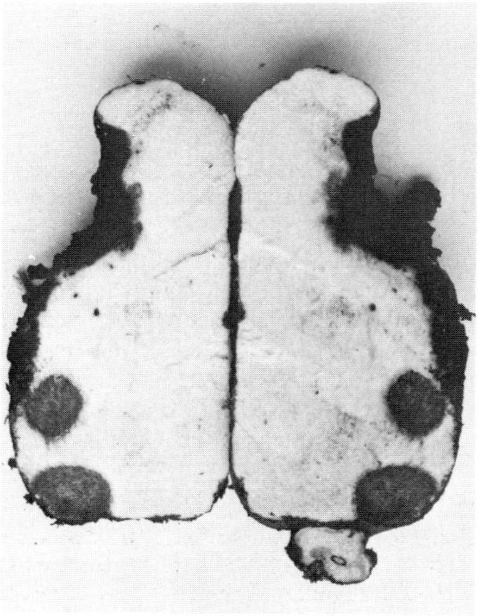

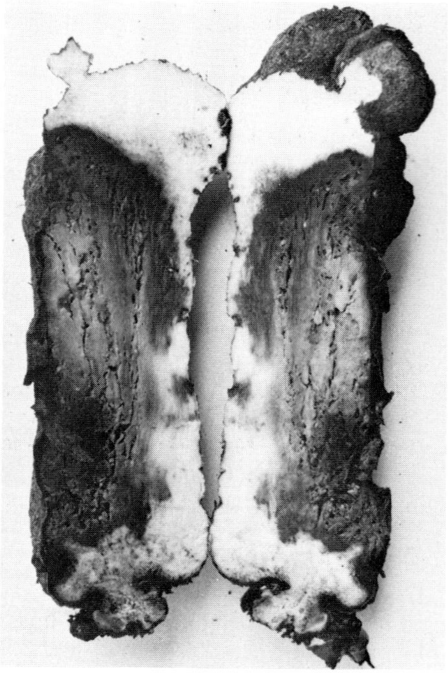

Figs. 70 and 71. Tubers cut open to show wet rot caused by species of *Botryodiplodia* and *Fusarium*. The photograph on the right shows a tuber also infested by the nematode *Scutellonema bradys* and showing dry rot symptoms. (E. Terry)

Control

Tubers used for planting should be carefully selected. Rotting often occurs when cut tubers have been planted without time being allowed for cut surfaces to dry out, as this facilitates entry of the pathogen. The incidence of infection can be greatly reduced by control measures such as treating the cut surfaces with Bordeaux mixture or coating with lime, chalk or wood ashes. Crop rotation also helps to reduce the disease incidence.

Dipping the tubers in systemic fungicides such as benomyl or thiabendazole has also been found to be effective. In one trial a five minute dip in 3000 ppm benomyl followed by dusting the cut ends with 0.3% benomyl in talcum powder effectively controlled tuber rot. In another trial 30 minute dips in 1000 ppm benomyl or thiabendazole gave 80 to 90% control of pre-harvest rotting in the field.

Rhizopus nodosus

Tuber rot

Distribution

Ivory Coast, India.

Symptoms

R. nodosus causes a soft rot which particularly affects *D. esculenta*; only weak infection occurs in *D. alata.* The tubers shrink by up to three quarters of their volume and the skin becomes wrinkled.

When the tubers are cut, browning of the tissues is observed and thick light coloured liquid can be squeezed out of infected tissues. The infection spreads throughout the tuber, only the vascular tissue remaining uninfected. Eventually the tuber becomes mummified as the infected tissues dry out. A dry rot is also attributed to *Rhizopus nodosus*. *R. nodosus* has been shown to be a wound parasite.

Control

The damage caused by *R. nodosus* is not often economically important.

Careful selection of planting material and crop rotation should keep the disease in check. Control measures, such as fungicide dips, applied for other tuber rots are also effective against *R. nodosus.* As nematodes facilitate entry of this pathogen a careful watch should be kept on their population and appropriate controls applied where necessary (see page 166).

Fusarium oxysporum, F. solani, F. sp.

Tuber rot

Distribution

China, India, Ivory Coast, Japan, Nigeria and Puerto Rico.

Symptoms

Nematode infestation facilitates the entry of *Fusarium* spp. into the tubers. Infected tissues develop a brown coloured dry rot. Numerous formae speciales of *F. oxysporum* and *F. solani* have been recorded as pathogens of yams. The symptoms caused are generally similar although there is some variation in the severity and the colour of the infected tissues. Often infection is not noticed until after harvest or until the tubers are cut open. Infected tubers may feel lighter than healthy ones and lack firmness when pressed. If the tubers are cut transversely zonations of infection can be seen. Brown dry rot caused by *Fusarium* occurs centrally and there may be a blackish zone round this where there is considerable nematode damage and secondary bacterial infection.

Control

As for *Botryodiplodia theobromae.* Also avoid planting yams in nematode-infested soil or use suitable measures to control nematodes (see page 166).

Minor root diseases

Armillariella mellea (*CMI Descriptions of Pathogenic Fungi and Bacteria* No. 321) has been recorded as the causal agent of a root rot of yams in Tanzania.

Penicillium sclerotigenum is recorded as a causal agent of yam tuber rot in China, Japan and India. Control measures, where necessary, are the same as for *Botryodiplodia theobromae.*

Rosellinia bunodes (*CMI Descriptions of Pathogenic Fungi and Bacteria* No. 351) is recorded as a causal agent of yam tuber rot in Jamaica.

Sphaerostilbe repens (*CMI Descriptions of Pathogenic Fungi and Bacteria* No. 391) is recorded as a causal agent of tuber rot in the Ivory Coast. This rot only develops appreciably in waterlogged or alkaline soils.

VIRUS DISEASES

Virus diseases of the mosaic type have been reported from West Africa and the Caribbean but they probably occur worldwide.

Mosaic type symptoms have been observed on *D. rotundata, D. cayenensis, D. alata* and *D. esculenta* and the wild *D. minutiflora* in West Africa. The disease is best known on *D. rotundata.* There is considerable variation in the susceptibility of cultivars within any species.

A wide range of symptoms may be observed. Leaves become mottled with dark green and chlorotic patches; vein clearing frequently occurs. Leaves may be distorted at the edges. When the infection is severe the whole plant appears stunted and may be less than a metre in height. Many shoots are produced both at the base and from the lateral buds, giving the plant a bushy appearance. The effect on yield is slight when symptoms develop near the end of the growing season. Virus-like disease symptoms on *D. rotundata* in Nigeria include green vein-banding (Fig. 72), shoestring bushiness, leaf distortion, leaf mottling and stunting. Studies have shown that the mottling and green vein-banding symptoms could be readily induced by mechanical and vector (*Aphis gossypii*) transmission. Non-persistent insect transmission is indicated.

Work with yam mosaic in Puerto Rico has shown that the strain in this area is readily transmitted by sap inoculation. It could also be transmitted by tuber grafting but no insect vector was positively identified.

The incidence of mosaic in both West Africa and Puerto Rico appears to be very localised. It is not yet known whether the Puerto Rican virus is the same as the Nigerian one. Although there are no confirmed reports of insect transmission of the Puerto Rican virus, *Aphis gossypii* is suspected to be a vector.

Fig. 72. Green vein banding symptoms on *Dioscorea rotundata*. (E. Terry)

In an investigation of yam virus of *D. alata, D. cayenensis, D. esculenta, D. rotundata* and *D. trifida* in the Caribbean, symptoms of mild mottle, severe mosaic vein banding, vein clearing and leaf distortion were observed. The various symptoms appeared to be caused by a single virus but may be caused by different strains of one virus or different closely related viruses. Forty eight percent of plants sampled showed virus-like symptoms of one kind or another. Fifty percent of samples examined by electron microscopy contained virus-like particles. These particles were long flexuous rods with an average length of 770 nm (range 600–950 nm). They were found in all five yam species. It appears that this virus is the same as the yam mosaic virus reported from Puerto Rico and Barbados by earlier authors. Symptoms of green vein banding were described on *D. composita* and *D. floribunda* in Puerto Rico. It would appear that this is caused by the same virus.

An internal brown spotting (Fig. 73) is associated with virus infection of tubers from Barbados. The affected tubers develop hard brown nodules in the flesh sometimes surrounded by necrotic areas, and produce foliage with mosaic symptoms. Flexuous rod virus-like particles can be isolated from both healthy and diseased plants. Bacilliform virus-like particles were also observed and seem to be a pre-requisite for the development of internal brown spot. The bacilliform particles seemed indistinguish-

TABLE 14. YAM VIRUS DISEASES

Symptoms	Transmission	Virus particles	Host	Area where the work was carried out	Author
Dark green bands bordering main veins with yellowish green in between	Mechanical. Tuber grafting. Possibly *Aphis gossypii*	Not described	*D. composita*, *D. floribunda*	Puerto Rico	Ruppel *et al.* (1966)
Yellow and dark green patches on the leaves with slight distortion of the lamina. Infected plants smaller. Mosaic	Sap inoculation — carried in tubers	Not described	*D. rotundata*	Puerto Rico	Adsuar (1955)
Internal brown spot. Mosaic of leaves	–	Flexuous virus-like particles and bacilliform particles	*D. alata*	Barbados	Harrison and Roberts (1973)
Dioscorea latent	Mechanical – Not aphids	Particles 395–445 nm long 2nd virus — not described	*D. floribunda*, *D. composita*	Puerto Rico	Waterworth *et al.* (1974)
Green banding	Aphids and mechanical	Pinwheel inclusions			
Severe symptoms of mosaic mottling, vein banding etc. Severe and mild symptoms and normal looking plants		Bundles of filamentous virus-like particles. Pinwheel cytoplasmic inclusions	*D.* spp	Caribbean	Mohamed (1976)
Internal brown spot and spot free	–	Bacilliform virus-like particles 125 nm × 25 nm	*D. alata*, *D. cayenensis*		
Spot free	–				
Numerous virus-like symptoms	–	Flexuous rod 770 nm long	*D.* spp	Caribbean	Mohamed and Mantell (1976)
Green vein banding and other virus like symptoms	Mechanical *A. gossypii*	Not described	*D. rotundata*	Nigeria	Terry (1976)
Mosaic and green banding	*Myzus persicae*	Pinwheel inclusions. Particles 395–445 nm. Particles 690–755 nm not observed in *D. floribunda* but transmitted by *M. persicae* to *D. composita*	*D. floribunda*	Puerto Rico	Lawson *et al.* (1973)

Fig. 73. Internal brown spot (R. H. Booth).

able from those of cocoa swollen shoot virus and it is thought that they belong to the same group.

Virus diseases of yams are not generally of any great economic importance; however, localised serious attacks can occur. The best way of preventing serious outbreaks is to select planting material from apparently disease-free plants. Seriously affected plants should be rogued.

Bibliography

ADSUAR, J. (1955). A mosaic disease of the yam *Dioscorea rotundata* in Puerto Rico. *Journal of Agriculture of the University of Puerto Rico* 39(2): 111–112.

AGRAWAL, S. B. and GUPTA, M. N. (1974). A new record of *Penicillium sclerotigenum* causing dry rot of 'Ratalu' *Dioscorea alata. Indian Phytopathology* 26(3): 577–578.

BAMMI, R. K., RAO, V. R., RANDHAWA, G. S. and SELVARAJ, Y. (1972). Control by benomyl of tuber rot in *Dioscorea floribunda. Plant Disease Reporter* 56(11): 990–991.

BAUDIN, P. (1956). Maladies parasitaires des ignames en Côte d'Ivoire. *Revue de Mycologie. Paris* 21(1): 87–111.

CHOUDHURY, S. (1946). A *Rhizoctonia* leaf blight of *Dioscorea. Current Science* 15(3): 81–82.

COURSEY, D. G. (1967). *Yams.* pp. 230. Longmans, London.

EKUNDAYO, J. A. and NAQVI, S. H. Z. (1972). Preharvest microbial rotting of yams (*Dioscorea* spp.) in Nigeria. *Transactions of the British Mycological Society* 58(1): 15–18.

FOURNET, J., DEGRAS, L., ARNOLIN, R. and JACQUA, G. (1975). [Experiments on yam anthracnose.] Essais relatifs a l'anthracnose de l'igname. *Nouvelles Agronomiques des Antilles et de la Guyane* 1(2): 115–122.

GRAHAM, K. M. (1971). *Plant Diseases of Fiji.* Overseas Research Publication No. 17. pp. 250. Ministry of Overseas Development, London.

HARRISON, B. D. and ROBERTS, I. M. (1973). Association of virus-like particles with internal brown spot of yam (*Dioscorea alata*). *Tropical Agriculture* 50(4): 335–340.

KOCH, W. (1974). Tropishe Knollenpflanzen and ihre Krankheiten. *Nachrichtenblatt für den Deutschen Pflanzenschutzdienst* (*Braunschweig*) 26: 129–135.

LAWSON, R. H., HEARON, S. S., SMITH, F. F. and KAHN, R. P. (1973). Electron microscopy and separation of viruses in *Dioscorea floribunda. Phytopathology* 63: 1435 (Abstract).

MACMILLIAN, R. T. Jr. (1974) Effective fungicides for control of yam anthracnose. *Phytopathology* 64(6): 769.

MATUE, T., OZAWA, M. and NAKAZAWA, T. (1973). Formae speciales of *Fusarium oxysporum* and *F. solani* causing 'brown rot' of fleshy root of *Dioscorea batatas* Decne. *Reports of the Tottori Mycological Institute* 10: 541–552.

MOHAMED, N. A. (1976). Virus-like particles and cytoplasmic inclusions associated with diseased *Dioscorea* spp. in the Eastern Caribbean. *Tropical Agriculture* (*Trinidad*) 53(4): 341–351.

MOHAMED, N. A. and MANTELL, S. H. (1976). Incidence of virus symptoms in yam (*Dioscorea* sp.) foliage in the Commonwealth Caribbean. *Tropical Agriculture* (*Trinidad*) 53(3): 255–261.

OGUNDANA, S. K. (1972). The control of soft rot of yams in Nigeria. *International Biodeterioration Bulletin* 8(2): 75–78.

PRASAD, N. and SINGH, R. D. (1960). Anthracnose disease of *Dioscorea alata. Current Science* 29(2): 66–67.

RUPPEL, E. G., DELPHIN, H. and MARTIN, F. W. (1966). Preliminary studies on a virus disease of a sapogenin-producing *Dioscorea* spp. in Puerto Rico. *Journal of Agriculture of the University of Puerto Rico* 50: 151–157.

SINGH, R. D. and PRASAD, N. (1966). Efficacy of different fungicides for control of anthracnose of *Dioscorea alata. Plant Disease Reporter* 50(6): 385–7.

SINGH, R. H. and PRASAD, N. (1967). Epidemiological studies on the anthracnose of *Dioscorea alata* L. (Ratalu) — yam. *Indian Phytopathology* 20(3): 226–36.

TERRY, E. R. (1977). Incidence, symptomatology and transmission of a *Dioscorea* spp. virus in Nigeria. *Proceedings of the Fourth Symposium of the International Society for Tropical Root Crops,* Cali, Colombia, 1976. pp. 170–174. International Development Research Centre, Ottawa, Canada.

THAUNG, M. M. (1974). Two new rusts from Burma. *Transactions of the British Mycological Society* 62(1): 218–222.

WAITT, A. W. (1963). Yams, *Dioscorea* species. *Field Crop Abstracts* 16(3): 145–157.

WATERWORTH, H. E., LAWSON, R. H. and KAHN, R. P. (1974). Purification, electron microscopy, and serology of *Dioscorea* latent virus. *Journal of Agriculture of the University of Puerto Rico* 58(3): 351–357.

NEMATODES*

Scutellonema bradys

The yam nematode

Distribution

Brazil, Cuba, India, Ivory Coast, Jamaica, Nigeria, Puerto Rico, Senegal and Togo.

Symptoms

The yam nematode is a migratory parasite of both yam tubers and roots. It invades tubers through or alongside roots, through growing tips and also via cracks in the skin. Eggs are mainly laid in the plant tissues where they hatch and develop into adults. Populations in excess of 100,000 nematodes can be found in infested tubers and all active stages of the life cycle are infective.

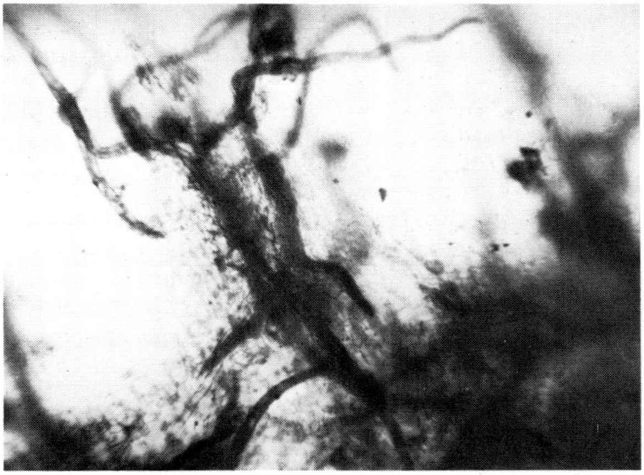

Fig. 74. *Scutellonema bradys* nematodes in dry rot tissue of a *D. rotundata* tuber. (J. Bridge)

S. bradys (Fig. 74) causes a disease of tubers commonly known as dry rot. The nematodes are normally restricted to the outer 1–2 cm of tuber in the peridermal and sub-peridermal layers. They feed intracellularly in the tissues causing cell destruction and necrosis. In the early stages of infection, small yellow lesions can be observed just

*This chapter was written by Dr J. Bridge, ODM Liaison Officer in plant nematology.

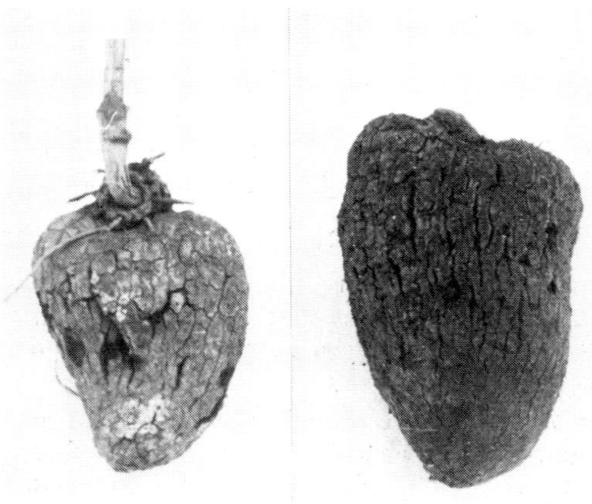

Fig. 75. Dry rot of *D. rotundata* caused by *S. bradys* showing cracking of tuber skin. (J. Bridge)

Fig. 76. Tubers of *D. alata* showing flaking off of the skin due to severe dry rot caused by *S. bradys*. (J. Bridge)

below the tuber skin; these become dark brown and eventually form a continuous dark, dry rot layer just beneath the surface in heavily infested tubers. Outwardly the infected tubers have cracks in the skin, which are often deep (Fig. 75), can be malformed, or the skin flakes off exposing the dark brown, dry rot tissue underneath (Fig. 76). Populations of *S. bradys* less than 1,000 nematodes/50 g of tuber do not cause external symptoms in *D. rotundata*.

S. bradys can cause dry rot symptoms in the absence of other organisms, but nematode damage does predispose the tubers to secondary pathogens, particularly fungi and bacteria, which give rise to a wet rot of internal tuber tissues. Nematode damage to the tubers results in a considerable reduction in the edible portion and marketable value of the tubers. Storage losses are most severe because the nematodes continue feeding in the stored yams associated with an increase in both dry and wet rot and can lead to complete destruction of stored tubers. Moisture loss is significantly greater from diseased tubers during storage and those harvested late from dry soil.

Many *Dioscorea* spp. are hosts for *S. bradys,* including *D. alata, D. bulbifera, D. cayenensis, D. dumetorum, D. esculenta* and *D. rotundata.* Many other crops also support populations of the nematode, such as cowpea, okra, melon, sorghum and pigeon pea. *S. bradys* is the major nematode pest in West Africa but not elsewhere in the main yam growing regions.

Pratylenchus coffeae

Lesion nematode

Distribution

P. coffeae occurs worldwide as a serious pest of many important crops, but has been found as a pest of yams only in Puerto Rico, Jamaica and British Solomon Islands.

Symptoms

P. coffeae occurs in yam roots and tubers as a migratory endoparasite. Large numbers of the nematode can be found in both soil and roots, but highest populations are found in tubers especially in the oldest portions adjacent to the vines.

Symptoms of damage to yam tubers by *P. coffeae* are similar to those caused by *S. bradys*. The nematodes are restricted to the outer layers of tuber tissue; they cause severe necrosis or dry rot, cracking of the skin, and are associated with complete deterioration of tubers. Both *P. coffeae* and *S. bradys* parasitise yams in Puerto Rico, but the former is the dominant species and causes most of the damage to tubers, mainly during storage. *P. coffeae* populations can increase considerably in yams stored at 22–31°C but remain low at temperatures less than 13°C.

The only other clearly identified *Pratylenchus* species found in yams is *P. brachyurus* which has been reported from Ivory Coast, Nigeria and Guatemala.

Meloidogyne spp.

Root-knot nematodes

Distribution

Worldwide, probably in all yam growing regions.

Symptoms

Root-knot nematodes are common and serious pests of tropical crops. Yams have been found infested by *M. incognita, M. javanica, M. arenaria* and *M. hapla*. The second stage juveniles invade roots and tubers and migrate to a feeding site normally adjacent to the vascular system. The adult female is swollen and remains stationary; eggs are laid either in the tissues of the plant, or on the surface of the roots and tubers. Feeding induces galling of the tissues around the nematode's body and causes a characteristic knotting when the nematode occurs in roots. Tubers infested by *Meloidogyne* spp. have an uneven, knobbly appearance (Fig. 77), often with small, necrotic spots in internal tissues surrounding individual nematodes. Yam seedlings infested with root-knot nematodes can be severely stunted and chlorotic; very young seedlings may be killed by severe infestation. Rotting of tubers has been reported, but most evidence suggests that *Meloidogyne* spp. do not survive within tubers during storage.

Fig. 77. Yam tubers damaged by the root-knot nematode *Meloidogyne incognita.* The specimen second from the right is severely damaged by *Scutellonema bradys.* (R. O. Ogbuji)

M. incognita has been found widespread on both *D. rotundata* and *D. alata* in parts of Nigeria; *M. javanica* and *M. arenaria* also cause galling of yam tissues in the country. In Guatemala, *M. incognita* and *M. arenaria* are the main nematode species infesting the sapogenin producing yams; *D. composita, D. floribunda* and *D. spiculiflora.*

Other nematodes

One species of nematode, *Aphelenchoides besseyi,* occurs in both the foliage and tubers of *D. trifida* in the French West Indies. *Trichodorus porosus* is associated with a black scurf-like syndrome of unknown aetiology on seedlings of *D. opposita* in Japan.

A large number of other plant parasitic nematodes have been reported in or around yam roots but information on their importance is not available, however, *Rotylenchulus reniformis,* the reniform nematode, is commonly found in both yam roots and tubers.

Control

Nematodes will decline in the absence of host plants and a fallow period has been suggested to control *S. bradys* in *D. alata* tubers. Attempts at controlling yam nematodes in field soil have proved relatively unsuccessful. Application of D-D at 448 *l*/ha has reduced the number of infected tubers from 88% to 32% in Puerto Rico. In Nigeria, use of nematicides has produced varying results, however both D-D and DBCP have given higher yields of 'clean' tubers when applied at rates of 84–253 *l*/ha and 28–84 kg/ha respectively.

Many of the nematodes are disseminated in infected tubers and the use of nematode-free tubers for propagation will effectively reduce the incidence of damage. Both hot water treatment and chemical dips have given some degree of control in tubers. Immersion of *D. rotundata* tubers in water at 46–52° for 15–30 min have given good control of *P. coffeae* in Puerto Rico. Hot water treatment at 51°C for 30 min also kills most root-knot nematodes in *D. floribunda* tubers. Good control of *M. incognita* has been reported in both *D. floribunda* and *D. composita* without damage to tubers by treating at 45°C for 30 min. In Jamaica, treatments at 50°C for varying lengths of time reduced populations of *P. coffeae* but tubers of *D. rotundata* and *D. cayenensis* did suffer severe physiological damage. Treatments at 50°C and 55°C for 40 min has given over a 98% kill of *S. bradys* without damage to *D. rotundata* tubers, but sprouting is delayed if tubers are treated soon after dormancy has broken, and, if treated later than this, they are inclined to rot.

Immersion of *D. rotundata* tubers in DBCP and fensulfothion at 1250 ppm for 15–30 min effectively reduces nematode populations (mainly *P. coffeae*) present to a depth of 6 mm in the tuber tissues in Puerto Rico.

Gamma radiation has been tried unsuccessfully to control *S. bradys* in stored tubers of *D. rotundata.*

Bibliography

ACOSTA, N. and AYALA, A. (1975). Pathogenicity of *Pratylenchus coffeae, Scutellonema bradys, Meloidogyne incognita* and *Rotylenchulus reniformis* on *Dioscorea rotundata. Journal of Nematology* 7: 1–6.

ACOSTA, N. and AYALA, A. (1976). Effects of *Pratylenchus coffeae* and *Scutellonema bradys* alone and in combination on Guinea yam (*Dioscorea rotundata*). *Journal of Nematology* 8: 315–317.

ACOSTA, N. and AYALA, A. (1976). Hot water and chemical dips for nematode control in tubers of *Dioscorea rotundata. Journal of Agriculture, University of Puerto Rico* 60(3): 395–402.

ADESIYAN, S. O. (1976). Host range studies of the yam nematode, *Scutellonema bradys. Nematropica* 6: 60–63.

ADESIYAN, S. O., ODIHIRIN, R. A. and ADENIJI, M. O. (1975). Histopathology studies of the yam tuber (*Dioscorea rotundata* Poir) infected with *Scutellonema bradys* (Steiner & LeHew). *International Biodeterioration Bulletin* 11: 48–55.

ADESIYAN, S. O., ODIHIRIN, R. A. and ADENIJI, M. O. (1975). Economic losses caused by the yam nematode, *Scutellonema bradys,* in Nigeria. *Plant Disease Reporter* 59: 477–480.

BAUDIN, P. (1956). Les maladies parasitaires des ignames en Cóte d'Ivoire. *Revue de Mycologie,* Paris, Supplement 21, No. 2: 87–111.

BRIDGE, J. (1972). Nematode problems with yams (*Dioscorea* spp.) in Nigeria. *PANS* 18: 89–91.

BRIDGE, J. (1973). Nematodes as pests of yams in Nigeria. *Mededelingen Faculteit Landbouwwetenschappen, Rijks Universiteit, Gent* 38: 841–852.

BRIDGE, J. (1975). Hot water treatment to control plant parasitic nematodes of tropical crops. *Mededelingen Faculteit Landbouwwetenschappen, Rijks Universiteit, Gent* 40: 249–259.

BRUHN, C. and KOCH, W. (1962). Beobachtungen Über den Befall mittelamerikanischer Dioscoreenarten durch *Meloidogyne incognita* (Kofoid & White). *Nematologica* 8: 307–317.

CAVENESS, F. E. (1967). Shadehouse host ranges of some Nigerian nematodes. *Plant Disease Reporter* 51: 115–119.

GOODEY, T. (1935). Observations on a nematode disease of yams. *Journal of Helminthology* 13: 173–190.

KERMARREC, A. and ANAIS, A. (1974). Presence of *Aphelenchoides besseyi* in foliage and tubers of yam (*Dioscorea trifida*) in the French West Indies. *Nematropica* 4: 2–3.

MIÉGE, J. (1957). Influence de quelques caractères des tubercules semences sur la levée et le rendement des ignames cultivées. *Journal d'agriculture tropicale et de botanique appliquée* 4: 315–342.

NISHIZAWA, T. (1973). Pathogenicity of *Trichodorus porosus* to Chinese yam. *Japanese Journal of Nematology* 3: 33–37.

SCHIEBER, E. and LASSMAN, D. K. (1961). Root-knot nematode on *Dioscorea* in Guatemala. *Plant Disease Reporter* 45: 981.

SMIT, J. J. (1967). Nematodes. In *Yams*. D. G. Coursey, pp. 115–120. Longmans, London.

STEINER, G. and BUHRER, E. M. (1934). Disease symptoms produced by *Anguillulina pratensis* in yams. *Phytopathology* 24: 164–165.

THOMPSON, A. K., BEEN, B. O. and PERKINS, C. (1973). Nematodes in stored yams. *Experimental Agriculture* 9: 281–286.

WEST, J. (1934). Dry rot of yams. *Bulletin of the Imperial Institute* 32: 449–450.

INSECT PESTS

Introduction

Yams in West Africa are attacked by various tuber borers, which may seriously reduce the yield. Elsewhere the tubers and growing plants are damaged by scale insects and mealybugs which prevent the development of the sett. In addition to the insects described below other pests such as soft scales, grasshoppers, various beetles, lepidoteran larvae and crickets have been recorded occasionally causing damage to yam foliage and stems. In the absence of any other recommendations for specific pests not mentioned in this manual the farmer may try dimethoate for the sucking pests, carbaryl for Lepidoptera and beetle control and fenitrothion for grasshopper or cricket control.

Gymnogryllus lucens (Orthoptera: Gryllidae)

Cricket

Distribution

West Africa.

Symptoms

Tubers and vines are eaten by this insect, young plants may die as a result of the attack.

Control

Chemical

Soil insecticides as used for *Heteroligus meles* will control *G. lucens.*

Heteroligus meles (Coleoptera: Dynastidae)

Yam beetle

Distribution

Tropical Africa.

Symptoms

The planting setts are eaten by the adult beetles just after the rains have begun and in extreme cases the plants may wilt and die. Adult *H. meles* are similar to *Papuana* spp. and *Ligyrus* sp. in appearance (Figs. 98 and 99). They are 23–33 mm long, dark brown to black in colour, with two conspicuous tubercles on the head almost forming a transverse ridge. Punctures on the elytra form four longitudinal bands. Plants with vines taller than 1.25 m are more attractive to the beetles than shorter ones. Further attacks may occur just before harvest in October (in Nigeria) when the beetles feed again, causing serious damage (holes and lesions 1–2 cm in diameter), reducing the tuber quality to such a degree that they are unsaleable and predisposed to decay.

Life cycle

Heteroligus meles breeds in damp soil close to rivers. Eggs are laid in the soil in November and December (in Nigeria). The larvae are typical dynastid larvae (similar to Fig. 98), creamy white to grey in colour with brown head capsules. They feed on dead organic matter in the soil whilst in the first instar. During the second and third instars the larvae feed on grass roots in addition to organic matter. Third instar larvae moult to a quiescent pre-pupal stage that moults to become the pupa. Under laboratory conditions development from egg to adult takes 22–24 weeks. The beetles fly on a feeding migration to yam growing areas soon after the beginning of the rains. During the feeding migration the beetles can be attracted to light traps. They feed on the yam setts or newly sprouted plants for 4–5 months. From August to September (in Nigeria) they hardly feed at all. When the yam crop has developed they begin to feed voraciously before departing on their breeding migration, during November and December, back to the swampy areas.

Control

Chemical

Dusting the planting setts, the planting holes or soil surface at planting or during the feeding migration period with 7 g of 2.5% aldrin or 0.5% γ-BHC will control this pest. It has been shown that the residues found in the tubers after these pesticides have been used are acceptable, they are particularly low if the soil, rather than the planting sett is treated. Endosulfan dust (5%) is also recommended. Yam peel from treated tubers should not be eaten by humans or animals particularly if aldrin has been used.

Heteroligus appius (Coleoptera: Dynastidae)

Lesser yam beetle

Distribution

Southern Nigeria.

Symptoms

Tubers are damaged by adult and larval *H. appius* feeding on them and yields are reduced.

Life cycle

Adult *Heteroligus appius* are similar in appearance to *H. meles* but slightly smaller, 21.0–22.5 mm long. The adults migrate to yam gardens to breed. Eggs are laid close to the yam and the larvae feed on the tubers and roots before pupating in the soil.

A hibernation migration to wetter areas, similar to those in which *H. meles* breeds, takes place in November and December.

There are several other 'yam beetles' among them are *Prionoryctes caniculus* and *Prionoryctes rufopiceus* recorded in the southern rain forest vegetative zone close to the Niger.

Control

See *H. meles.*

Palaeopus costicollis (Coleoptera: Curculionidae)

Yam weevil

Distribution

Cuba and Jamaica.

Symptoms

Tuber setts used for planting may fail to develop if heavily attacked. The tubers, especially the head ends, are tunnelled by small (4 mm long) dull black red-legged weevils and their legless larvae. The market value of the crop which is subsequently attacked is further reduced by localised rotting around the damaged parts of the tubers. This insect has also been recorded from sweet potato.

Control

Cultural

Crop rotation should be practised and clean planting material used.

Chemical

The treatments recommended for the control of *H. meles* should control this weevil.

Aspidiella hartii (Hemiptera: Diaspididae)

Yam scale

Distribution

Widespread in the Old World tropics and Pacific (CIE Distribution Map 217).

Symptoms

Adult and nymphal scale insects (whitish yellow in colour) are present on the tubers and sometimes on the aerial portions of the plant, the plants do not thrive. Stored tubers may be thickly covered in scales (Fig. 78) and shrivel. The problem is perpetuated by the use of infested tubers as planting material.

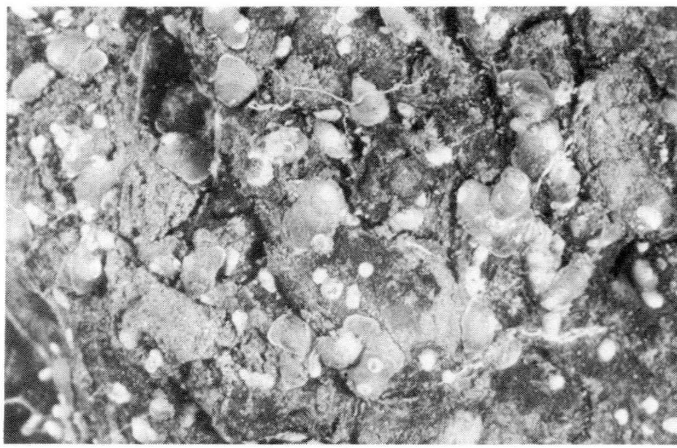

Fig. 78. *Aspidiella hartii* the yam scale on a yam tuber. (G. Swaine (1971) *Agricultural Zoology in Fiji*)

Control

Cultural

Aspidiella hartii and other scale insects on the tubers may be controlled by using clean planting material. Yams should not be planted after ginger, as this crop also supports the pest.

Chemical

Dipping the tubers in 0.3% aldrin before planting, in a mixture of diazinon 20% e.c. (56 ml) plus white oil (425 ml) in 13.5 *l* water, or in 1:20 mixture of DDT dust and wood ash has been recommended. Fumigation with carbon disulphide has been suggested.

Senoclidia purpurata (Hymenoptera: Tenthredinidae)

Sawfly

Distribution

Papua New Guinea.

Symptoms

Defoliation by numbers of larvae feeding together, often in rows, on large leaves, beginning at the apex of the leaf and moving backwards towards the leaf stalk. In serious infestations adult sawflies (Fig. 79) are commonly seen flying around the garden.

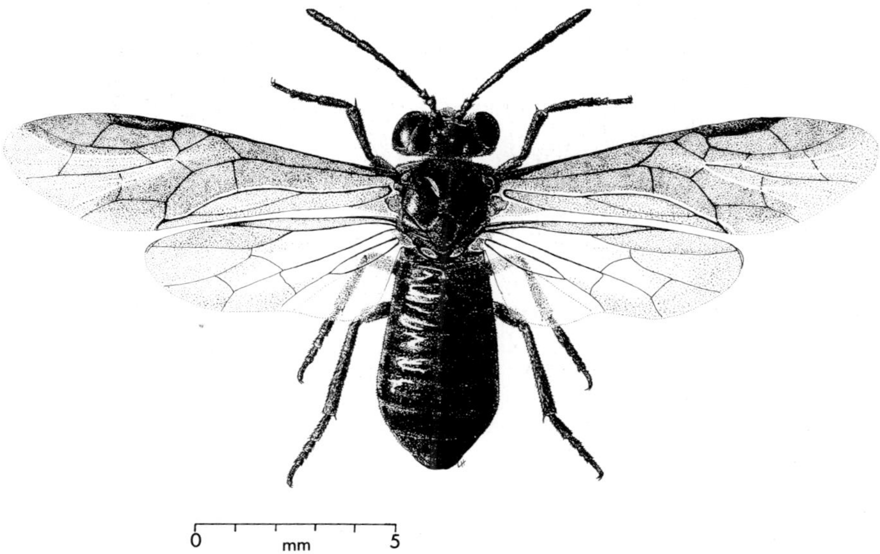

Fig. 79. *Senoclidia purpurata,* a sawfly reported damaging yams in Papua New Guinea. (L.H.)

Life cycle

The life cycle of this species is not recorded, the Tenthredinidae usually lay eggs in young shoots and leaves. Larvae are similar in appearance to Lepidoptera larvae but with more than five pairs of abdominal limbs. Pupation takes place in the soil.

Control

Outbreaks of this pest are unusual and specific control methods have not been recommended. Carbaryl applied at the manufacturers recommended rate will kill sawfly larvae.

Lilioceris livida (Coleoptera: Chrysomelidae)

Distribution

Throughout West Africa and east to the Kilimanjaro area.

Symptoms

Larvae and adults feed on the young leaves and shoots. These larvae cover themselves with their own excreta.

Life cycle

Pink elongate eggs 2 mm long are laid on the underside of the leaves in groups of about six. As they develop they darken to reddish brown. The larvae are thick and fleshy. Adults are basically reddish-brown to black, with light yellow to brown elytra on which are dark brown to black spots.

Control

Cultural

Hand picking and destruction of larvae have been recommended.

Chemical

Carbaryl (w.p. 0.2% a.i.), dichlorvos (e.c. 0.075% a.i.) or endosulfan (dust, e.c. or w.p. 0.075% a.i.) may be used.

Geococcus coffeae **(Hemiptera: Pseudococcidae)**
Phenacoccus gossypii
Planococcus citri
Planococcus dioscoreae

Mealybugs

Distribution

Geococcus coffeae: Pantropical, (CIE Distribution Map 285).

Phenacoccus gossypii: French Guiana, Guadeloupe and Peru.

Planococcus citri: Most tropical and subtropical areas and in glasshouses in Europe and North America. (CIE Distribution Map 43).

Planococcus dioscoreae: Papua New Guinea.

Symptoms

All these mealybugs have been reported feeding on the roots and/or tubers of yams. Adult female mealybugs are small, segmented, whitish-grey, wingless insects covered in a mealy wax secretion. Some also feed on the stem and leaves. The mealybugs are usually noticed when the yams are harvested, and can multiply often to considerable numbers causing shrivelling on stored yams. If affected yams are planted they may not sprout.

Life cycle

The female adults lay eggs from which nymphs emerge and develop through several immature stages. Any male nymphs present develop wing buds and eventually fully developed wings.

Control

Cultural

Only clean healthy tubers should be used as planting material.

Chemical

Dipping the planting material in aldrin or BHC solution or dusting in DDT dust have been recommended. The white oil-diazinon mixture recommended for *Aspidiella hartii* control may be expected to control mealybugs also. Any aerial parts of the plant affected may be dusted or sprayed with BHC, DDT, malathion or omethoate. Local recommendations as to the concentration of the insecticide used should be followed.

Bibliography

ANON. (1951). Yam weevil. *The Farmer* Jan–Feb–March: 26.

COURSEY, D. G. (1967). *Yams.* pp. 230. Longmans, London.

FENNAH, R. H. (1947). *The insect pests of food crops in the Lesser Antilles.* pp. 207. Department of Agriculture, Antigua, BWI.

MARSHALL, G. A. G. (1918). A new weevil pest of sweet potatoes in Jamaica. *Bulletin of Entomological Research* 8: 269–272.

PANIS, A., FERRAN, A. and TORREGROSSA, J-P. (1974). Les cochenilles des cultures vivrières et fourragères aux Antilles et en Guyane Française (Homoptera, Coccoidea) *Revue de Zoologie Agricole et de Pathologie Végétale* 73(1): 22–27.

SINGH, S. R. (undated). *Common root crop pests of Fiji and their control.* Ministry of Agriculture, Fisheries and Forests, Fiji.

SMITH, J. G. (1965). Yam pests. *The Ghana Farmer* 9(1): 5–7 and 29.

SZENT-IVANY, J. J. H. (1974). Insect pests of yam in Papua New Guinea. *Folia Entomologica Hungarica* 27(2): 205–210.

TAYLOR, T. A. (1964). Studies on Nigerian yam beetles: I Systematic notes on the common Nigerian yam beetle species (Coleoptera: Dynastidae) *Journal of the West African Science Association* 8(2): 180–189.

TAYLOR, T. A. (1964). Studies on Nigerian yam beetles: II Bionomics and control. *Journal of the West African Science Association* 9(1):13–31.

TAYLOR, T. A. (1969). Insecticide residues in yams treated against yam beetle (*Heteroligus* sp.) infestation in Nigeria. *Journal of the West African Science Association* 13(2): 139–145.

TEDDER, M. M. and TEDDER, J. L. O. (1974). *Yams; A description of their cultivation on Guadalcanal in the Solomon Islands.* Technical Paper No. 169. South Pacific Commission, Noumea, New Caledonia.

WILLIAMS, D. J. (1960). A new species of *Planococcus* Ferris (Coccoidea: Homoptera) on yams in New Guinea. *Papua and New Guinea Agricultural Journal* 13(1): 39–40.

TARO

DISEASES

Introduction

In many areas taro is grown with no serious disease problems; however, in other areas diseases cause serious losses and have occasionally resulted in taro cultivation being abandoned. Corm rots cause considerable damage but virus diseases and leaf blight are the most devastating in some areas. Strict quarantine measures should therefore be enforced to prevent the spread of these diseases. Crop rotation is important in the control of corm rots.

The distributions quoted for some diseases may be incomplete owing to the lack of available records.

FUNGAL DISEASES

Phytophthora colocasiae

Leaf blight

Distribution

This disease occurs widely in India, Indonesia, Malaysia, Sarawak and the Pacific. It has also been recorded from parts of Africa and the Caribbean. (CMI Distribution Map 466).

Symptoms

Phytophthora leaf blight is the most destructive leaf disease of taro. Lesions are initially small, dark and round but rapidly enlarge to 2.5–5 cm in diameter and become purplish to brownish in colour. Drops of a clear liquid exude from the spots, and turn yellow, orange or purple when dry. There is usually a chlorotic halo round the spots. As the disease progresses the spots coalesce and have characteristic rings of yellow or brown colour (Fig. 80). Eventually the whole leaf may be affected and may die.

Development and spread

Taro leaf blight is common in wetland taro and is favoured by humid, cloudy conditions and poor soil fertility. Losses of up to 50% have been reported. Spores are produced on the leaf spots and are readily spread by rain. Spores have been shown to remain viable in the laboratory for up to three months but in the field spores only survived for up to 16 days.

Fig. 80. Taro leaf with multiple infection of *Phytophthora colocasiae.* Note downward development of subsidiary lesions from primary infection on the left lobe. (Jackson and Gollifer (1975), *PANS* 21(1): 45–53)

Control

Cultural

Wide spacing of taro plants minimises spread of the disease due to rain splashing. Wide spacing together with interplanting and open sides to the taro field help improve the air flow and thus keep the humidity down. Farmers should avoid walking through the field when it is wet as this encourages the spread of spores. Where possible the taro should be cultivated on higher land. Crop remains should be burnt after harvest.

In cases of mild infection of young plants (up to two to three months) leaf rogueing can be effective in controlling or delaying the build-up of the disease. Rogued leaves should be burnt and care should be taken to avoid spreading spores from diseased leaves to healthy ones.

Resistant varieties

Generally the search for taro cultivars resistant to leaf blight has been unsuccessful. However, resistant cultivars are available in India. In some areas where *P. colocasiae* is severe on taro, the resistant or immune *Xanthosoma* spp. are sometimes grown instead.

Chemical

Various fungicides have been evaluated for control of taro leaf blight. Tribasic copper sulphate and copper oxychloride have been used successfully at intervals of seven to fifteen days depending on weather conditions and the severity of the disease (Fig. 81). Captafol was found to be extremely phytotoxic although it gives good control. Zineb and mancozeb have been shown to give effective control. The best control with mancozeb was obtained with sprays of 2.24 or 4.48 kg/ha at intervals of 5–7 days. It has been suggested that copper oxychloride sprays and phytosanitary measures could be used together in a practical control programme. Strict hygiene should be observed at planting and infected leaves should be removed twice a week for 80–90 days after planting, or until a dense canopy is formed. Great care should be taken to avoid spreading spores to healthy leaves. After rogueing four to six applications of copper oxychloride are then made before harvest.

Fig. 81. Good control of *Phytophthora colocasiae* leaf infection with copper oxychloride applied at 14 day intervals. (Jackson and Gollifer (1975), *PANS* 21(1): 45–53)

Cladosporium colocasiae

Brown leaf spot

Distribution

Borneo, Ethiopia, Fiji, Formosa, Guinea, Hawaii, Hong Kong, India, Nepal, New Caledonia, New Guinea, Nigeria, Pakistan, Sarawak, Solomon Is. and Tahiti.

Symptoms

Superficial dark reddish-brown spots 5–10 mm in diameter develop on both sides of the leaves. The spots are diffuse in appearance. The disease occurs in wetland and upland taro and is particularly severe on older leaves.

Control

The disease is not of economic importance and no specific control measures are recommended. Fungicide sprays applied for blight control will also control brown leaf spot.

Phyllosticta colocasiicola

Shothole leaf spot

Distribution

Fiji, Puerto Rico and Samoa.

Symptoms

Shothole leaf spot can be confused with *Phytophthora* leaf blight. The first symptoms are brown spots with pale yellowish-white haloes. The spots enlarge to 2.5–5 cm in diameter and the middles fall out giving the characteristic 'shot hole' appearance. The holes may coalesce giving the leaves a ragged appearance. The disease is more prevalent in outer leaves.

Development and spread

Spores are produced by pycnidia on the surface of the spots and are spread by rain. Infection occurs mostly during the cool season especially in shaded or weedy sites.

Control

No specific control measures are recommended. Good husbandry and regular weeding will help to minimise the disease.

Minor leaf diseases

Pseudocercospora colocasiae

Cercosporidium sp.

Myrothecium roridum (*CMI Descriptions of Pathogenic Fungi and Bacteria* No. 253)

Pleosphaerulina sp.

Leptosphaerulina trifolii

Johnstonia colocasiae

Phytophthora palmivora

Alternaria tenuissima

Macrophomina phaseolina (*CMI Descriptions of Pathogenic Fungi and Bacteria* No. 275)

The above pathogens cause leaf diseases of taro but are not of economic importance or are of local importance only. In older leaves *Thanatephorus cucumeris* (*CMI Descriptions of Pathogenic Fungi and Bacteria* No. 406) infects the petiole base resulting in early collapse but damage is rarely significant.

Pythium spp.

Phytophthora spp.

Corm rot

Distribution

India, Pacific area.

Causal agent

Pythium spp. are the most frequently isolated pathogens from corms with rot symptoms. Several *Pythium* spp. have been recorded in Hawaii and the New Hebrides. (In the Solomon Islands *P. myriotylum* (*CMI Descriptions of Pathogenic Fungi and Bacteria* No. 118) has been regularly isolated from decaying roots, often associated with *P. splendens* (*CMI Descriptions of Pathogenic Fungi and Bacteria* No. 120) and other *Pythium* spp.).

Phytophthora cinnamomi (*CMI Descriptions of Pathogenic Fungi and Bacteria* No. 113) and *Phytophthora nicotianae* var. *parasitica* are the main causes of corm rot in Fiji. In some cases the pathogenicity of the isolates has not been established. Species of both *Pythium* and *Phytophthora* are soil-borne and it is probable that some of the species causing corm rot are found wherever taro is grown. There is certainly a considerable amount of secondary infection following invasion by these fungi which makes isolation of the primary causal agent difficult. However, descriptions of the symptoms of corm rot are essentially similar and so are the control methods.

Symptoms

If young plants are attacked they may be killed without producing leaves or they may become severely stunted. In plants over three months old the first symptoms observed are wilting, chlorosis and stunting which may be followed by collapse of the whole plant (Fig. 82). The rotting of the corm usually starts at the base and progresses upwards so that eventually the whole corm may be affected. The rotted tissue is firm and may be whitish-yellow to grey and blue or dark purple; healthy and diseased tissues are clearly divided. In later stages the diseased tissue crumbles like cheese and has an offensive smell. There is often secondary infection with *Erwinia chrysanthemi* which causes a strong-smelling white soft rot.

Roots decay in association with the corms. In some cases there is little damage to the corm but many roots are killed leaving the plant loose in the ground (Fig. 83). Rotting may also spread from the main corm to lateral suckers.

Fig. 82. Taro plants at three months from planting showing above ground symptoms following root infection by *Pythium myriotylum*. (Jackson and Gollifer (1975), *PANS* 21(1): 45–53)

Fig. 83. Roots of taro showing various stages of root destruction by *Pythium myriotylum*. (Jackson and Gollifer (1975), *PANS* 21(1): 45–53)

Development and spread

These fungi are encouraged by plentiful moisture, warm temperature and poor soil conditions. Any physical damage to taro plants predisposes them to infection. If taro is left in the field after maturity it causes a build-up of the disease. These fungi survive between crops in crop residues. Depending on conditions, damage may be confined to a few corms or may be widespread and cause up to 100% loss. Damage is particularly severe in wetland areas and poorly drained soil.

Control

Cultural

Taro should be harvested as soon as possible after it is mature and crop debris should be removed from the fields. Diseased plants should be rogued from the growing crop taking care to remove as much of the corm and decayed roots as possible. (N.B. This is not a practical control measure when fields are heavily contaminated with root-rotting fungi.). If a field is badly infected with the disease taro should not be cultivated for two or three seasons and the soil should be encouraged to dry out by deep cultivation. In irrigated wetland taro water should be kept flowing and not be allowed to stagnate. Dryland taro should be planted in well drained ground. Healthy planting material should always be used.

Resistant varieties

Screening of varieties has shown that there is considerable variation in crop tolerance. Therefore it is best to select a variety which is suitable for local conditions and which has tolerance to corm rot where necessary.

Chemical

Dipping of planting material in fungicides such as copper sulphate or captan has been recommended. Soil drenches with these fungicides have also been found to be effective.

Corticium rolfsii

Sclerotium rot

Distribution

Widespread (CMI Distribution Map 311)

Symptoms

Corticium rolfsii (*CMI Descriptions of Pathogenic Fungi and Bacteria* No. 410) causes a rot of dying petioles and a corm rot. The rotted tissue is brown and soft but not watery and tends to be stringy. Characteristic white mycelial growth and numerous dark brown sclerotia which resemble cabbage seed are observed. The disease may be more severe in taro that is over mature.

Development and spread

Corticium rolfsii is a common pathogen and attacks many crops; sources of inoculum are therefore readily available. The disease is more common in upland taro.

Control

Prompt harvesting of taro and crop rotation help to avoid the disease. Infected plants should be removed and burnt. Soil drenches with quintozene or dicloran have been recommended for control.

Fusarium oxysporum

Soft rot

A soft rot of corms caused by *Fusarium oxysporum* (*CMI Descriptions of Pathogenic Fungi and Bacteria* No. 211) is sometimes observed at harvest. The diseased corms have a whitish-grey, spongy soft rot which usually has a definite brown margin round it. In cases of severe corm rotting foliage becomes wilted and the plant may collapse. Rotted corms have an offensive smell.

Minor root and corm diseases

Fusarium solani (*CMI Descriptions of Pathogenic Fungi and Bacteria* No. 29) dry rot of corms

Endoconidiophora sp. black rot of corms

Phytophthora colocasiae corm rot of eddoe

Diplodia sp. black rot of corms

The above corm diseases have been recorded but are not of widespread economic importance. Post-harvest rots are discussed in the chapter on storage (pp. 41–42).

VIRUS DISEASES

Dasheen mosaic virus

Vectors *Myzus persicae, Aphis gossypii, Pentalonia nigronervosa*

Distribution

Egypt, Hawaii, India, Japan, Papua New Guinea, Puerto Rico, Trinidad, USA, Venezuela.

Symptoms

Dasheen mosaic virus (DMV) has long flexuous rod-shaped particles measuring about 75 nm in length. The presence of the virus does not always result in the production of symptoms or symptoms may occur only on a few leaves of the plant. Symptom expression may vary from one taro variety to another. DMV infection results in a feathery mosaic pattern adjacent to the veins. Yellow-green streaking of the petioles may also be observed in *Xanthosoma* spp. In severely affected plants petioles are short and leaves are dwarfed and fail to unfurl properly. DMV has been recorded on taro, eddoe and *Xanthosoma* spp. and occurs commonly in other araceous genera e.g. *Caladium, Dieffenbachia, Philodendron* and *Amorphophallus.*

Control

Only healthy planting material should be used.

Hot water treatment of corms is not effective for controlling DMV. It is possible to obtain healthy planting material by meristem tip culture. In the future it may be possible to produce pathogen free certified stock as a result of multiplication of material produced by meristem tip culture. This would be done under carefully controlled conditions at research stations and the certified stock would be released to growers.

Rogueing is unlikely to be successful as symptoms are not always present. There will be no information on the effect of DMV on yield until yields of plants known to be virus free are compared with those of DMV infected ones.

Vectors

DMV is transmitted non-persistently by the aphids *Myzus persicae, Aphis gossypii* and *Pentalonia nigronervosa.* This virus is also sap inoculable. The chapter on insects should be consulted for information on vector control.

Alomae and Bobone

Vectors *Tarophagus proserpina* and *Planococcus citri*

Distribution

Cook Islands, Papua New Guinea, Solomon Islands.

Symptoms

Two viruses are found in taro affected by Alomae and Bobone. One of these has large bacilliform particles (280 × 55 nm) and the other has small bacilliform particles (140 × 30 nm).

Both types of virus particle are found in plants with Alomae whereas in plants with Bobone only the large particle is present.

In the Solomon Islands chromosome counts have shown that in a majority of cases taro susceptible to Alomae have a chromosome number of 42. These are generally high yielding taro and known locally as 'male' cultivars. Smaller taro known as 'female' are susceptible to Bobone and have a chromosome number of 28. However, there are exceptions to this situation. Taro sent from Hawaii, Fiji and New Hebrides also considered to have a chromosome number of 28 were found susceptible to Alomae when tested in the Solomon Islands.

Alomae is a systemic necrosis which leads to death of the plant. The first symptom is a feathery mosaic. Young leaves fail to open properly and have a crinkled appearance. The leaves become abnormally thickened and the veins prominent. Later in the development of the disease the leaves do not open and begin to die. The necrosis spreads from the tips of the leaves and down the petioles and finally the whole plant dies.

Symptoms of Bobone are similar to those of Alomae but milder. Affected plants tend to be more stunted and the leaves are twisted and curled. The foliage does not become necrotic but remains green and the plants eventually recover.

Recent studies of taro viruses have shown that either particle may be found alone in 'male' or 'female' cultivars. In most of these cases the plants recover within four to eight weeks of the first appearance of symptoms. When the large particle occurs alone in 'male' plants mild Bobone symptoms of leaf distortions alone are observed. The small particle alone in 'male' or 'female' plants causes stunting, chlorosis and crinkling of the leaves, which often fail to open properly. In a very few cases 'female' plants exhibiting Bobone-type symptoms have been found to contain both types of particle but they always recovered from the disease. Symptoms may be confused by the additional presence of dasheen mosaic virus.

Examination of taro showing symptoms other than those of Alomae and Bobone has shown that the infected plants usually contain either the large or small bacilliform particles or the flexuous rods (dasheen mosaic virus) or a combination of these. Table 15 gives examples of the symptoms observed and the viruses associated with them.

TABLE 15. OCCURRENCE OF VIRUS PARTICLES IN EDIBLE AROID MATERIAL FROM THE SOUTHWEST PACIFIC

Location	Host*	Symptom	Viruses detected†		
			Lb	Sb	Fr
Malaita	*C. esculenta* (F)	Mild mosaic, reflexing and chlorosis of leaves; stunting, then recovery within 4 weeks (Fig. 84)	–	+	±
Malaita	*C. esculenta* (F)	Leaves chlorotic, puckered, cupped and often torn; usually recover but sometimes die. May develop from or into a Bobone symptom (Fig. 85)	–	+	±
Malaita	*C. esculenta* (M)	Leaf reflexing, chlorosis and stunting shortly after planting. May recover or die (Fig. 86)	–	+	±
Malaita	*C. esculenta* (M)	Young leaves are chlorotic and fail to unfurl after planting. Plants usually die (Fig. 87)	–	+	±
Malaita	*C. esculenta* (M)	Scarred, wrinkled areas on occasional leaf laminae; petiole galls sometimes present. Plant always recovers (Fig. 88)	+	–	±
Malaita	*Alocasia* sp.	Laminal puckerings and chlorosis	–	–	–
Choiseul	*C. esculenta*	Reflexed laminae with mild yellow vein-clearing. Inspreading marginal chlorosis and necrosis on old leaves. Persistent	–	+	–
Santa Ysabel	*C. esculenta*	Severe leaf chlorosis and stunting of plant which always dies	–	+	–
Guadalcanal	*C. esculenta*	Remain stunted after planting	–	+	–
Fiji (Viti Levu)	*C. esculenta*	Stunting or marginal necrosis of laminae	–	+	–
Fiji (Viti Levu)	*C. esculenta*	Laminal mosaic	–	–	+

Location	Host*	Symptom	Viruses detected†		
			Lb	Sb	Fr
Tongatapu	*C. esculenta*	Laminal mosaic	–	–	+
Tongatapu	*Xanthosoma* sp.	Laminal mosaic	–	–	+
Raratonga (Cook Is.)	*C. esculenta*	Laminal mosaic	–	–	+
Raratonga	*Xanthosoma* sp.	Laminal chlorosis and puckering	–	+	+
Samoa	*C. esculenta*	Petiole enations	–	–	–
Bututitari (Gilbert Is.)	*Cyrtosperma* sp.	Laminal mosaic	–	–	+
Singapore	*C. esculenta*	Interveinal chlorosis	–	–	–
Singapore	*Alocasia* sp.	Leaf chlorosis and puckering	–	–	–
Kenya	*C. esculenta* (3 cvs.)	None	–	–	–
Ghana	*C. esculenta* (3 cvs.)	None	–	–	–
Peru	*C. esculenta* (3 cvs.)	None	–	–	–

* F = 'female'; M = 'male'.
† Lb = Large bacilliform; Sb = Small bacilliform; Fr = Flexuous rod.
± present in some samples but not in others.
From Gollifer *et al.* (1977), *PANS* 23(2): 171–177.

Transmission

The larger bacilliform particle is transmitted persistently by the planthopper, *Tarophagus proserpina.* The smaller bacilliform particle is transmitted by the mealybugs *Planococcus citri* and *Pseudococcus longispinus.* Alomae may be apparently 'caused' by *T. proserpina* transmitting the larger bacilliform particle to 'male' plants already infected with the smaller particle.

Control

Areas which are free from Alomae and Bobone should impose strict quarantine measures to prevent the importation of infested and infected material.

Where these diseases occur infected plants should be rogued and burnt and planting material should be carefully selected. Although the 'female' cultivars can survive infection, corm yield losses of about 25% still occur, so it is still advisable to rogue infected plants as they provide a reservoir of virus particles.

Figs. 84 and 85. Female taro plants showing two types of symptom apparently caused by infection with the small bacilliform particle (see Table 15). (Gollifer *et al.* (1977), *PANS* 23(2): 171–177)

Figs. 86 and 87. Male taro plants showing two types of symptom apparently caused by infection with the small bacilliform particle (see Table 15). (Gollifer *et al.* (1977), *PANS* 23(2): 171–177)

Fig. 88. Male taro plant showing scarring and wrinkling of the leaf caused by infection with the large bacilliform particle (see Table 15). (Gollifer *et al.* (1977), *PANS* 23(2): 171–177)

Some work has been done on biological control of *Tarophagus proserpina* (see p. 197), however, this is unlikely to be an effective method of controlling these virus diseases. Chemical control (see p. 200) of *T. proserpina* is not effective when populations are high.

Bibliography

ALCONERO, R. (1972). Hot water treatments of corms of *Xanthosoma* spp. infected with dasheen mosaic virus. *Plant Disease Reporter* 56(4): 320–321.

ALCONERO, R. and ZETTLER, F. W. (1971). Virus infections of *Colocasia* and *Xanthosoma* in Puerto Rico. *Plant Disease Reporter* 55(6): 506–508.

BERGQUIST, R. R. (1972). Efficacy of fungicides for control of *Phytophthora* leaf blight of taro. *Annals of Botany* 36(145): 281–287.

BERGQUIST, R. R. (1974). Effect of fungicide rate, spray interval, timing of spray application and precipitation in relation to control of *Phytophthora* leaf blight of taro. *Annals of Botany* 38(154): 213–221.

DEBROT, E. A. and ORDOSGOITTI, A. (1974). Dasheen mosaic virus infection of *Colocasia* and *Xanthosoma* in Venezuela. *Plant Disease Reporter* 58(11): 1032–1034.

GOLLIFER, D. E. and BOOTH, R. H. (1973). Storage losses of taro corms in the British Solomon Islands Protectorate. *Annals of Applied Biology* 73(3): 349–356.

GOLLIFER, D. E. and BROWN, J. F. (1972). Virus diseases of *Colocasia esculenta* in the British Solomon Islands. *Plant Disease Reporter* 56(7): 597–599.

GOLLIFER, D. E. and BROWN, J. F. (1973). Corm rot of *Colocasia esculenta* caused by *Fusarium oxysporum*. *Plant Disease Reporter* 57(8): 701–703.

GOLLIFER, D. E. and BROWN, J. F. (1974). *Phytophthora* leaf blight of *Colocasia esculenta* in the British Solomon Islands. *Papua New Guinea Agricultural Journal* 25(1/2): 6–11.

GOLLIFER, D. E., JACKSON, G. V. H., DABEK, A. J., PLUMB, R. T. and MAY, Y. Y. (1977). The occurrence and transmission of viruses of edible aroids in the Solomon Islands and the South West Pacific. *PANS* 23(2): 171–177.

GRAHAM, K. (1965). Dalo diseases. *Fiji Farmer* 1(4): 54–57.

GRAHAM, K. M. (1971). *Plant Diseases of Fiji.* Overseas Research Publication No. 17. pp. 250. Ministry of Overseas Development, London.

GROVER, R. K. and PRASHER, R. D. (1968). *Rhizoctonia* leaf rot of *Colocasia* in India. *Plant Protection Bulletin, FAO* 16(1): 6–7.

HICKS, P. G. (1967). Resistance of *Colocasia esculenta* to leaf blight caused by *Phytophthora colocasiae. Papua and New Guinea Agricultural Journal* 19(1): 1–4.

JACKSON, G. V. H. (1977). *Taro leaf blight.* South Pacific Commission Advisory Leaflet No. 3. pp. 4.

JACKSON, G. V. H., BALL, E. A. and ARDITTI, J. (1977). Tissue culture of taro, *Colocasia esculenta* (L.) Schott. *Journal of Horticultural Science* 52:373–382.

JACKSON, G. V. H. and GOLLIFER, D. E. (1975). Disease and pest problems of taro (*Colocasia esculenta* L. Schott) in the British Solomon Islands. *PANS* 21(1): 45–53.

JACKSON, G. V. H. and GOLLIFER, D. E. (1975). Storage rots of taro (*Colocasia esculenta*) in the British Solomon Islands. *Annals of Applied Biology* 80(2): 217–230.

KENTEN, R. H. and WOODS, R. D. (1973). Viruses of *Colocasia esculenta* and *Xanthosoma sagittifolium. PANS* 19(1): 38–41.

KOCH, W. (1974). Tropische Knollenpflanzen und ihre Krankheiten. *Nachrichtenblatt für den Deutschen Pflanzenschutzdienst* (Braunschweig) 26: 129–135.

KULKARNI, S. N. and SHARMA, O. P. (1975). Corm rot of *Colocasia antiquorum* Schott. due to *Phytophthora colocasiae* Sacc. *JNKVV Research Journal* (*Jawakarlal Nehru Krishi Vishma Vidyalaya*) 9(1/2): 70.

O'CONNOR, B. A. (1969). *Exotic Plant Diseases.* South Pacific Commission, Noumea, New Caledonia.

PLUCKNETT, D. L., DELAPENA, R. S. and OBRERO, F. (1970). Taro (*Colocasia esculenta*). *Field Crop Abstracts* 23(4): 413–426.

SOLANKURE, R. T. and RAO, V. G. (1973). Alternaria leaf spot of taro from India. *Indian Phytopathology* 1972, 25(3): 457–459.

TRUJILLO, E. E. and AZAGAKI, M. (1964). Taro blight and its control. *Hawaii Farm Science* 13(4): 11–13.

UMABALA, K. and RAMARAO, P. (1972). Leaf blight of Colocasia caused by *Phytophthora palmivora. Indian Journal of Mycology and Plant Pathology* 2(2): 187–188.

YEN, D. E. and WHEELER, J. M. (1968). Introduction of taro into the Pacific: the indications of the chromosome numbers. *Ethnology* 7: 259–267.

ZETTLER, F. W., FOXE, F. J., HARTMAN, R. D., EDWARDSON, J. R. and CHRISTIE, R. G. (1970). Filamentous viruses infesting dasheen and other araceous plants. *Phytopathology* 60: 983–987.

NEMATODES*

Meloidogyne javanica, M. incognita

Root-knot nematodes

Distribution

To date they have been reported on *Colocasia* only from Fiji, Niue (Pacific Island), India and Trinidad.

Symptoms

In India, *M. javanica* causes severe losses to the crop. The foliage of nematode infested plants at first becomes yellow and then turns brown and ultimately dies back. The corms are deformed with blister-like swellings or galls varying in size from 2–15 mm. Rotting of corms associated with the nematode, occurs during storage. *M. incognita* is reported to cause only slight galling of roots and corms in Trinidad and the Pacific Islands.

Control

Application of fumigant and granular nematicides gave some control of *M. javanica* on *Colocasia* in India; best control was achieved with DBCP at 3.6 *l*/ha applied as a drench treatment.

Other nematodes

Large numbers of an undescribed species of *Hirschmanniella* have been found from roots and corms of taro plants examined from the Solomon Islands, where the nematodes are reported to cause severe reddening of internal corm tissues and render the crop inedible.

Other plant parasitic nematodes found in or around roots of *Colocasia* include *Criconemoides* spp., *Helicotylenchus* spp., *Longidorus siddiqi, Rotylenchus* sp., *Tylenchus* spp. and *Xiphinema insigne.*

There are very few reports of nematodes on *Xanthosoma* and little indication that they cause serious damage to the crop.

---

*This chapter was written by Dr J. Bridge, ODM Liaison Officer in plant nematology.

Rotylenchulus reniformis occurs in abundance on *Xanthosoma* in Trinidad associated with small root lesions. Other nematodes found are *Meloidogyne* sp., *Helicotylenchus* sp., *Tylenchorhynchus* sp. and *Criconemoides* sp. In Tonga, *Meloidogyne* sp. causes some galling of both roots and tubers.

There are no reports of control measures specifically against nematodes of *Xanthosoma.*

Bibliography

BRAITHWAITE, C. W. D. (1972). *Colocasia esculenta,* a new host of *Meloidogyne incognita* in Trinidad. *Plant Disease Reporter* 56: 618.

BRAITHWAITE, C. W. D. (1972). Preliminary studies on plant-parasitic nematodes associated with selected root crops at the University of the West Indies. *Plant Disease Reporter* 56: 1077–1079.

NIRULA, K. K. (1959). Root-knot nematode on *Colocasia. Current Science,* Bangalore 28: 125–126.

SRIVASTAVA, A. S., GUPTA, R. L., SINGH, B. and RAM, S. (1969). Damage and control of *Meloidogyne javanica* (Treub) in *Colocasia* (*C. antiquorum*). *Labdev Journal of Science and Technology* 7B: 306–307.

SRIVASTAVA, A. S., VERMA, R. S. and PANDEY, R. C. (1971). Comparative efficacy of different nematicides tested against root-knot nematode infesting *Colocasia antiquorum. Labdev Journal of Science and Technology* 9B: 142–144.

INSECT PESTS

Introduction

The most important insect pests of taro are root borers. Where these are known to occur it is advisable to treat the planting hole with an insecticide (see the section on *Papuana* spp.). It is not usually necessary to apply insecticide to control leaf-feeding insects unless very young plants are being attacked, established taro can suffer more than 50% defoliation before a reduction in yield can be detected.

LEAF FEEDERS

Aphis gossypii (Hemiptera: Aphididae)

Melon aphid, cotton aphid

Distribution

Cosmopolitan but absent from the cooler parts of Asia and Canada (CIE Distribution Map 18).

Vector of dasheen mosaic virus (see p. 185).

Symptoms

Nymphs, winged and wingless adult aphids are seen feeding on leaves, causing speckling, leaf curl and a generally sickly appearance (Fig. 89). A sooty mould may develop on the honeydew secreted by the aphids and in addition ants are attracted to the honeydew.

Life cycle

Adults may give birth to two nymphs per day. Winged adults (all female) have a black head with two green bands, the abdomen is green with black cornicles. Wingless adults (also all female) are yellowish-green to black.

Control

It is not necessary, or economic, to apply insecticides to control this pest on taro unless very severe infestation on young leaves prevents the establishment of the plant. If taro is planted close to an alternative host of the insect, such as melon, cucumber or other curcubits or cotton, it may be thought necessary to spray the taro to prevent the insect spreading to more profitable crops.

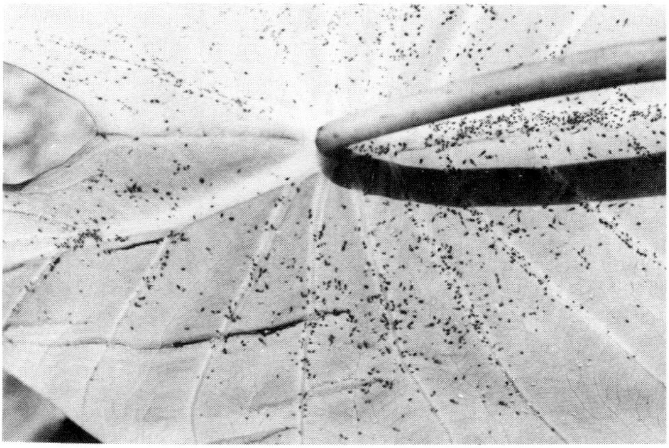

Fig 89. *Aphis gossypii* on the lower surface of a taro leaf. (G. Swaine (1971) *Agricultural Zoology in Fiji*)

Chemical

The recommendations in Table 16 are current in Fiji and the West Indies.

TABLE 16. RECOMMENDATIONS FOR CHEMICAL CONTROL OF *APHIS GOSSYPII*

Insecticide	Formulation	Active ingredient	Application
Carbaryl	w.p.	1,000 g/ha	In 450 *l* (high volume) or 112–168 *l* (low volume) of water
Diazinon	e.c.	11 ml	In 13.5 *l* water. 14 days minimum period to harvest
Dimethoate	e.c.	4.4 ml	In 13.5 *l* water. 14 days minimum period to harvest
Malathion		14.2 ml	In 13.6 *l* water. 7 days minimum period to harvest

Bemisia tabaci (Hemiptera: Aleurodidae)

Cassava, cotton or tobacco whitefly

Distribution

Recorded in most parts of the tropics and subtropics (CIE Distribution Map 284).

Symptoms

Adults (Fig. 90), nymphs and 'pupae' are found on the lower leaf surface. When infestation is heavy the leaves yellow, wilt and die, and growth of the taro may be stunted.

Fig. 90. *Bemisia tabaci,* a whitefly, on taro. (G. Swaine (1971) *Agricultural Zoology in Fiji*)

Life cycle

Yellow elliptical eggs, attached by filaments to the leaf, are laid on the undersurface of the leaves. The first instar nymphs emerge and feed. After the first moult they are fixed to one place on the leaf as yellow elliptical scale-like insects. The third moult gives rise to the 'puparium' which is less flat than the previous nymphal stages and the red eyes of the adult can be seen through the integument. Egg to adult development takes 2—3 weeks. The adult has a yellow body, white waxy broad tipped wings and is 1.0 mm long with a wingspan of about 2.0 mm.

Control

Chemical

In Fiji the materials in Table 17 have been recommended for the control of *B. tabaci* on taro.

TABLE 17. RECOMMENDATIONS FOR CHEMICAL CONTROL OF *BEMISIA TABACI*

Insecticide	Formulation	Active ingredient in 13.6 *l* water	Application
Carbaryl	w.p.	34.0 g	7 day minimum period before harvest
Malathion	w.p.	14.0 g	7 day minimum period before harvest
Trichlorphon	w.p.	23.0 g	7 day minimum period before harvest

Tarophagus proserpina (Hemiptera: Delphacidae)

Taro leafhopper

Distribution

Recorded in most South Pacific territories, Hawaiian Is., Indonesia, Malaysia, Philippines, Ryukyu Is. and Queensland (CIE Distribution Map 306).

Vector of the larger bacilliform virus particle responsible for Alomae and Bobone virus diseases in taro (see pp. 186–190).

Symptoms

A reddish crust of sap exudate produced when the insects puncture the plant during feeding is visible on the outside of the leaf stalks near the ground (Fig. 91). Plants which are heavily infested may be stunted in growth and may wilt. Brown and white adult and white nymphal leafhoppers (Fig. 92) can be seen.

Life cycle

The life cycle, egg to adult, takes about 18 days. The adult leafhoppers are about 4 mm long, dark brown in colour with a white patch on the thorax (Fig. 93). They have full sized or reduced wings. The eggs are laid in the midrib of the taro leaf and the immature leafhoppers which hatch from them are white.

Control

Biological

In Hawaii satisfactory control, reducing injury to taro to an unimportant level was achieved by the introduction of the mirid *Cyrtorhinus fulvus* in 1938.

Fig. 91. Leafhopper damage to taro. (G. Swaine (1971) *Agricultural Zoology in Fiji*)

Fig. 92. Nymphs of *Tarophagus proserpina* on taro. (G. Swaine (1971) *Agricultural Zoology in Fiji*)

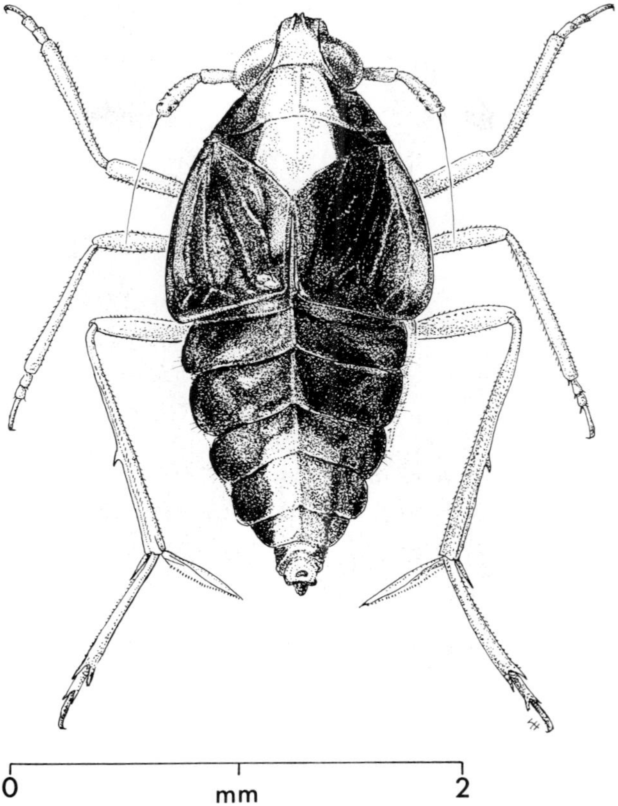

Fig. 93. The leafhopper *Tarophagus proserpina.* (L.H.)

Chemical

The recommendations in Table 18 have been found satisfactory in Fiji and the West Indies.

TABLE 18. RECOMMENDATIONS FOR THE CHEMICAL CONTROL OF *TAROPHAGUS PROSERPINA*

Insecticide	Formulation	Active ingredient in 13.6 *l* water	Application
Carbaryl	w.p.	34 g	7 days minimum period to harvest
Malathion	w.p.	28 g	7 days minimum period to harvest
	e.c.	28 ml	
Trichlorphon	w.p.	23 g	14 days minimum period to harvest

Hippotion celerio (Lepidoptera: Sphingidae)

Taro hornworm, taro hawkmoth, silver striped hawkmoth

Distribution

Throughout the Old World tropics, ranging into temperate regions, Australia and into the Pacific as far as Samoa.

Symptoms

Pieces are eaten out of the edges of leaves by hornworm larvae (Fig. 94) living on the underside. Severe defoliating attacks, including the destruction of leaf stalks to ground level, have been reported.

Life cycle

Eggs are laid on leaves and stems. The larvae, which emerge after three to four days, possess a straight, pink to black caudal horn. The body and head of older larvae are basically green (although brown forms do occur), the darker dorsal area is separated from the lower part of the body by a continuous creamy yellow lateral line from the fifth to caudal segment. Two 'eye' spots occur on the fourth and fifth segments. When about to pupate the larvae darken, the dorsal area becomes chocolate coloured and the ventral area tan. At this stage larvae are about 37 mm long with a head width of 3 mm and the legs are deep orange. After 10 to 15 days feeding the larvae pupate in the soil. The pupa is about 45 mm long, 10.5 mm wide at its widest point and tan colour with numerous dark brown dots. The adult, a hawkmoth 40 mm long with a wingspan of 70 mm, is basically brown in colour with a grey patch on the top of the head which extends to the thorax as a yellow and white line. The forewings are brown with a broad white band extending to the tip, and the hindwings have a red tinge near the base and are patterned with black.

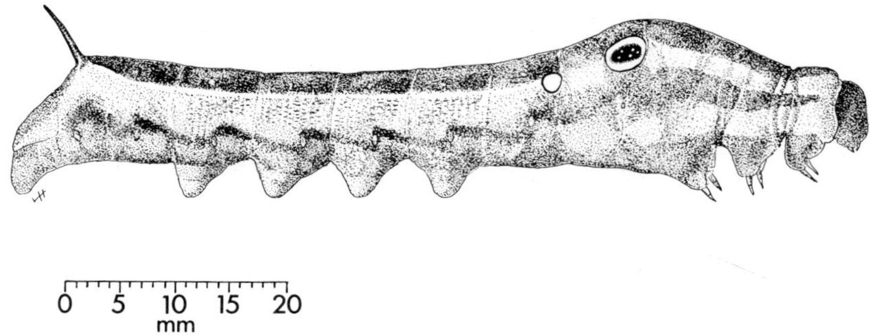

Fig. 94. Taro hornwom *Hippotion celerio.* (L.H.)

Control

Cultural

Hand picking of the larvae will often provide sufficient control.

Chemical

In Papua New Guinea a 0.1% a.i. solution of carbaryl applied with a knapsack sprayer is recommended if defoliation exceeds 50% of the leaf area.

Fig. 95. The cluster caterpillar *Spodoptera litura.* (G. Swaine (1971) *Agricultural Zoology in Fiji*)

Agrius convolvuli, the sweet potato hornworm (see p. 81) and *Spodoptera litura,* the taro cutworm or cluster caterpillar (Fig. 95) may also attack taro; they may be controlled by the methods recommended for *H. celerio.*

ROOT FEEDERS

Papuana **spp. (Coleoptera: Dynastidae)**

Taro beetles

Distribution

Papuana huebneri: Bismarck Archipelago, Gilbert Is., Moluccas Is., Papua New Guinea and the Solomon Is.

Papuana inermis: Solomon Is.

Papuana woodlarkiana: Bismarck Archipelago, Indonesia, Moluccas Is., Papua New Guinea, Queensland and Solomon Is.

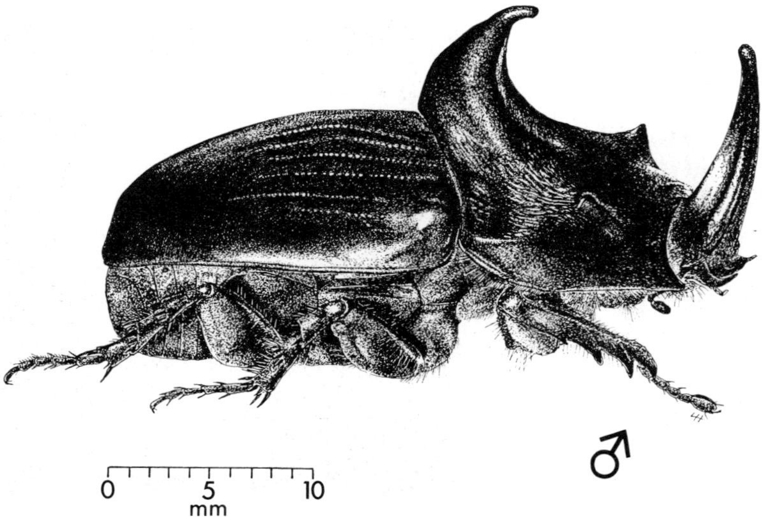

Fig. 96. Adult taro beetle, *Papuana woodlarkiana.* (L.H.)

Symptoms

Newly planted taro may wilt and die due to adult beetles (Fig. 96) boring in the corm and tunnelling up into the meristem tissue, via the crown used for planting material. The beetles are 20–27 mm long and dark brown to almost black in colour. Older plants wilt but usually recover. At harvest the corms are found to be riddled with smooth sided tunnels 10–15 mm in diameter (Fig. 97), they are then susceptible to rot if stored. Attacks seem to occur most frequently at the edge of the garden.

Life cycle

Eggs are laid in the soil about 50–150 mm below the surface near to a host plant, for example. *Saccharum spontaneum.* The larvae (Fig. 98) feed on organic matter and probably roots. After passing through three instars they pupate, from 200 mm to more than a metre below the soil surface. After pupation the adult emerges from the soil and locates a suitable food plant. Laboratory work has shown that *P. huebneri, P. woodlarkiana* and *P. inermis* take 122, 215 and 119 days respectively to develop from egg to adult.

Fig. 97. Taro damaged by *Papuana* sp. (B. Thistleton)

Fig. 98. *Papuana* sp. larva, probably *P. woodlarkiana.* (B. Thistleton)

Control

Chemical

In Papua New Guinea it is recommended that 6% γ-BHC granules or γ-BHC + carbaryl granules should be put into the planting holes at 0.84 kg a.i./ha. Alternatively 20% γ-BHC dust at 1 kg a.i./ha may be applied when planting and repeated 40 and 130 days after planting.

Ligyrus spp. (Coleoptera: Dynastidae)

Distribution

Ligyrus ebenus: Bolivia, Brazil, Colombia, Guyana, Mexico, Peru, Surinam, Trinidad and Venezuela.

Ligyrus cuniculus: Brazil, French Guiana and the West Indies.

Symptoms

The plants wilt due to both adults and larvae boring into the roots. Adults (Fig. 99) are dark brown to black, strongly convex dorsally, 25 mm long and 16 mm wide. If the attack is severe, death of the plants follows very quickly. This pest appears to be attracted to plants growing in wet or irrigated areas during dry weather.

Life cycle

Eggs are usually laid singly in soil and organic material. After 13 days the curved, six legged larvae hatch. After passing through four larval instars pupation takes place in the soil and lasts 14 days; the adults emerge at night.

Control

Cultural

In small gardens, wilting plants can be lifted and the insects killed by inserting wires into their tunnels. The taro can then be replanted.

Chemical

Chemical control methods have not been developed specifically for these pests as it is unlikely that they will be of economic importance. The larvae may be attacked by a parasitic wasp, *Campsomeris dorsata,* and also by a muscardine fungus. The possibilities of encouraging biological control should be studied.

Fig. 99. *Ligyrus ebenus* another beetle which feeds on taro roots. (L.H.)

Minor pests

Tetranychus marianae, and other spider mites, have been recorded as pests of taro, attacking the central roll of unexpanded leaves causing silvering, drying and distortion. Dicofol w.p. at 600 g a.i./ha is recommended, in the West Indies, for mite control on taro, with a minimum waiting period of 3 days if the leaves are to be used as spinach. Diazinon or demeton-methyl at the rate recommended by the manufacturers are recommended elsewhere.

Grasshoppers and bugs attack taro, sometimes seriously, carbaryl at the rate recommended for *Aphis gossypii* should control these insects.

The mealybugs *Planococcus citri* and *Pseudococcus longispinus* have been identified as vectors of a virus (see p. 188). Specific control methods for the control of these mealybugs as vectors of virus diseases have not been established, it is usual to rogue the diseased crop. Diazinon or malathion, used at the manufacturers recommended rate for mealybug control should prevent the spread of these insects.

Bibliography

ANON. (1976). Insects of traditional vegetables. Topic 11. *Entomology Conference, Papua New Guinea, 1976.* Unpublished report, Department of Primary Industry.

ANON. (1976). Vegetable pests. Topic 13. *Entomology conference, Papua New Guinea, 1976.* Unpublished report, Department of Primary Industry.

COMSTOCK, J. A. (1966). Lepidoptera of American Samoa with particular reference to biology and ecology. *Pacific Insects Monograph* 11: 1–74.

ENDRODI, VON S. (1971). Monographie der Dynastinae 4 Tribus Pentodontini (papuanische und pazifische Inselwelt) (Coleoptera: Lamellicornia: Melolonthidae) *Pacific Insects* 13(2): 243–320.

FENNAH, R. G. (1947). *The insect pests of food crops in the Lesser Antilles.* pp. 207. Department of Agriculture, Antigua, BWI.

GOLLIFER, D. E., JACKSON, G. V. H., DABEK, A. J., PLUMB, R. T. and MAY, Y. Y. (1977). The occurrence and transmission of viruses of edible aroids in the Solomon Islands and the southwest Pacific. *PANS* 23(2): 171–177.

JACKSON, G. V. H. and GOLLIFER (1975). Disease and pest problems of taro (*Colocasia esculenta* L. Schott) in the British Solomon Islands. *PANS* 21(1): 45–53.

MATSUMOTO, B. M. and NISHIDA, T. (1966). *Predator prey investigations on the taro leaf hopper and its egg predator. Technical Bulletin* No. 64. Hawaii Agriculture Experiment Station.

O'CONNOR, B. A. (1969). *Exotic plant pests and diseases.* pp. 424. South Pacific Commission, Noumea, New Caledonia.

PARASRAM, S. (1973). *Control of insect pests of some food crops in the Caribbean. Extension Bulletin* No. 7. pp. 29. University of the West Indies.

PERRY, C. H. (in press). A note on taro leafhopper and taro hawkmoth. *Papua and New Guinea Agricultural Journal.*

PERRY, C. H. (in prep.). The bionomics and control of taro beetle (*Papuana* spp.) in Papua New Guinea.

PERRY, C. H. (in press). The ecology and control of some taro pests in Papua New Guinea. *Proceedings 10th Waigani Seminar, Lae, May 1976.*

PLUCKNETT, D. L., DELAPENA, R. S. and OBRERO, F. (1970). Taro (*Colocasia esculenta*). *Field Crop Abstracts* 23(4): 413–426.

SINGH, S. R. (undated). *Common root crop pests of Fiji and their control.* pp. 10. Ministry of Agriculture, Fisheries and Forests, Fiji.

SMEE, L. (1965). Insect pests of sweet potato and taro in the territory of Papua and New Guinea: their habits and control. *Papua and New Guinea Agricultural Journal* 17(3): 99–101.

SWAINE, G. (1971). *Agricultural zoology in Fiji.* Overseas Research Publication No. 18. pp. 424. Foreign and Commonwealth Office, Overseas Development Administration, London.

UNDP/FAO (1976). *UNDP/FAO Survey of agricultural pests and diseases. Interim Report, March 1976. Part two Nematology, Entomology*. pp. 57. South Pacific Bureau for Economic Co-operation.

GLOSSARY

ACERVULUS (ACERVULI). A dense cushion-like mass of conidiophores and conidia formed by some fungi.

ANTICOAGULANT. A chemical which prevents blood from clotting.

APEX. The tip of a root or shoot.

BACILLIFORM. Shaped like a rod.

CONIDIOPHORE. A simple or branched hypha bearing one or more conidia.

CONDIDUM (CONIDIA). An asexually formed spore, produced by many species of fungi.

CORM. A rounded swollen underground stem.

CORMEL. Secondary corm produced from an old corm.

CORNICLES. A pair of tubes on the abdomen of some Hemiptera.

ECTOPARASITE. A parasite which lives on the outside of its host.

ENDOPARASITE. A parasite which lives inside its host.

ELYTRA. The hard outer wingcases of beetles.

FLEXUOUS. Full of bends and curves.

FRASS. Rubbish, including excrement, left by boring insects.

HYPHA (HYPHAE). One of the simple branched filaments of the mycelium of a fungus.

INSTAR. A stage in the development of an insect.

INTEGUMENT. The epidermis and cuticle of an insect.

LENTICEL. A pore in a stem or root.

MATRIX. Any substance on which a fungus lives.

MYCELIUM. A mass of hyphae.

NECROSIS. Death of plant cells, especially when resulting in the tissue becoming dark in colour; commonly a symptom of fungus infection.

PERIDERM. The outer corky layer of a plant stem.

PERITHECIUM (PERITHECIA). Flask-like organ bearing asci, which contain ascospores, of the Pyrenomycetes.

PETIOLE. The stalk of a leaf.

PUNCTIFORM. Shaped like a dot.

PYCNIDIUM (PYCNIDIA). Spore-producing organ of the Sphaeropsidales, often globose or flask-like.

RENIFORM. Kidney shaped.

RHIZOMORPH. Thread-like structure made of fungal hyphae.

RHIZOSPHERE. The region in the soil surrounding the root system of a plant.

ROOT-KNOT. Galls and swellings produced on roots as a result of nematode attack.

SCLEROTIUM (SCLEROTIA). A hard mass of fungal hyphae, usually black on the outside, crust like to globular and serving as a resting stage from which fructifications may be formed later.

SETT. A piece of tuber used as planting material.

SETA (SETAE). A stiff hair.

STELE. The central region of a stem or root, containing the vascular tissues.

STOMA (STOMATA). A small hole in the epidermis of a leaf or a young stem, together with two guard cells.

SUBERISED. Transformed into cork.

THORAX. The region of the insect body between the head and abdomen.

TUBER. A swollen underground stem.

TUBERCLE. A projection.

APPENDIX

CHECK LIST OF DISEASES

SWEET POTATO

Species	Synonymy	Common name
Albugo ipomoeae panduratae (Schw.) Swing.		White rust
Alternaria tenuissima (Kunze ex Pers.) Wilts.		Leaf spot
Botryodiplodia theobromae Pat.		
Botrytis sp.		Grey mould rot
Ceratocystis fimbriata Ellis & Halst.		Black rot
Cercospora ipomoeae Winter		*Cercospora* leaf spot
Diaporthe batatatis Harter & Field		Dry rot
Diplodia tubericola (Ell. & Ev.) Taub.		Java black rot
Elsinoe batatas Viégas & Jenk.		Scab
Fusarium oxysporum f. sp. *batatas* (Woll.) Snyd. & Hans		Stem rot
Fusarium spp.		Surface/end rot
Macrophomina phaseolina (Tassi) Goid.		Charcoal rot
Monilochaetes infuscans Ell. & Halst. ex Harter		Scurf
Mucor racemosus Fres.		*Mucor* rot
Penicillium sp.		Blue mould rot
Phaeoisariopsis bataticola (Cif. & Bruner) M. B. Ellis	*Cercospora bataticola* Cif. & Bruner	*Cercospora* leaf spot
Phyllosticta batatas (Thüm.) Cke		*Phyllosticta* leaf spot
Phymatotrichopsis omnivora (Duggar) Hennebert	*Phymatotrichum omnivorum* Dug.	Root rot
Plenodomus destruens Harter		Foot rot
Pseudocercospora timorensis (Cke) Deighton	*Cercospora timorensis* Cke *Cercospora batatae* Zimm.	*Cercospora* leaf spot

Species	Synonymy	Common name
Pythium spp.		Ring rot
Rhizopus spp.		Soft rot
Septoria bataticola Taub.		*Septoria* leaf spot
Streptomyces ipomoeae (Pers. & Mart.) Waks. & Hen.		Soil rot

CASSAVA

Species	Synonymy	Common name
Agrobacterium sp.		Stem gall
Armillariella mellea (Vahl ex Fries) Karst.	*Armillaria mellea* (Vahl ex Fries) Quélet	Root rot
Aspergillus spp.		Post-harvest decay, root rot
Bacillus spp.		Post-harvest decay, root rot
Botryodiplodia theobromae Pat.		Stem rot
Botryodiplodia spp.		Post-harvest decay
Cercospora viscosae Muller and Chupp		*Cercospora* leaf blight
Cercosporidium henningsii (Allesch.) Deighton	*Cercospora henningsii* Allesch. *C. cassava* Ell. & Ev *C. manihotis* P. Henn.	Brown leaf spot
Clitocybe tabescens (Scopoli ex Fries) Bresadola		Root rot
Corticium rolfsii Curzi	*Sclerotium rolfsii* Sacc.	Root rot
Corticium salmonicolor Berks. & Br.		Pink disease
Corynebacterium spp.		Post-harvest decay
Diplodia spp.		Post-harvest decay
Drechslera rostrata (Drechsler) Richardson & Fraser		Leaf blight
Erwinia cassavae (Hansford) Burkholder 1948		Bacterial stem rot
Erwinia spp.		Post-harvest decay

Species	Synonymy	Common name
Fusarium spp.		Root rot, post-harvest decay
Glomerella cingulata (Stonem.) Spauld. & Schrenk		Stem rot
Glomerella manihotis (Sacc.) Petr.	*Colletotrichum manihotis* P. Henn.	Anthracnose
Helicobasidium compactum Boedijn		Root rot
Microsphaera euphorbiae Berk. and Curt.		Powdery mildew
Oidium manihotis P. Henn.		Cassava ash disease
Penicillium spp.		Post-harvest decay
Periconia sp.		Leaf spot
Phaeoramularia manihotis (Stev. & Solheim.) M.B. Ellis	*Cercospora caribaea* Cif.	White leaf spot
Phoma sp.	*Phyllosticta manihot* Speg.	Concentric-ring leaf spot
Phomopsis manihotis Swarup, Chauhan & Tripathi		Blight
Phyllosticta manihobae Viegas		Concentric-ring leaf spot
Phyllosticta manihoticola Sydow		
Phytophthora cryptogea Pethyb. & Laff.		Root rot
Phytophthora drechsleri Tucker		
Phytophthora erythroseptica Pethyb.		
Pseudophaeolus baudonii (Pat.) Ryv.	*Polyporus baudonii* Pat. *Pheolus manihotis* Heim	Root rot
Pythium spp.		Post-harvest decay, root rot
Rhizoctonia spp.		Post-harvest decay, root rot
Rhizopus spp.		Post-harvest decay, root rot
Rigidoporus lignosus (Klotzsch) Imazeki	*Fomes lignosus* (Klotzsch) Bresadola	White root disease (white thread disease)
Rosellinia necatrix Prill.		*Rosellinia* root rot
Sphaceloma manihoticola Bitanc. & Jenk.		Superelongation disease
Uromyces manihotis Henn		Rust
Uromyces spp.		Rusts

Species	Synonymy	Common name
Verticillium dahliae Kleb.		Vascular wilt
Xanthomonas cassavae Wiehe & Dowson		Bacterial leaf spot
Xanthomonas manihotis (Arthuad-Berthet & Bondar) Starr		Cassava bacterial blight

YAMS

Species	Synonymy	Common name
Agrobacterium tumefaciens (E.F. Smith and Townsend) Conn.		Crown gall
Armillariella mellea (Vahl ex Fr.) Karst		Root rot
Aspergillus niger van Tiegh		Tuber soft rot
Botryodiplodia theobromae Pat.	*Lasiodiplodia theobromae* (Pat.) Griff. et Maub.	Tuber soft rot
Botrytis sp.		Tuber soft rot
Cercospora brasiliensis Averna		Leaf spot
Cercospora carbonacea Miles		Leaf spot
Cercospora pachyderma Syd.		Leaf spot
Cercospora sp.		Leaf spot
Colletotrichum capsici (Syd.) Butler and Bisby		Leaf spot
Colletotrichum sp.		Tuber soft rot
Corticium rolfsii Curzi	*Sclerotium rolfsii* Sacc.	Leaf spot, neck rot
Corynebacterium sp.		Dry rot of tubers
Curvularia spp.		Leaf spot
Erwinia sp.		Watery rot of tubers
Fusarium oxysporum Schlecht.		Tuber rot
Fusarium solani (Mart.) Sacc.		Tuber rot
Fusarium spp.		Tuber rot
Gloeosporium pestis Massee		Leaf spot
Glomerella cingulata (Stonem.) Spauld. & Schr.	*Colletotrichum gloeosporioides* (Penz.) Sacc.	Anthracnose

Species	Synonymy	Common name
Goplana dioscoreae Cumm.		Rust
Hendersonula toruloidea Natrass		Tuber soft rot
Macrophomina phaseolina (Tassi) Goid.		Tuber soft rot
Penicillium sclerotigenum Yamomoto		Tuber soft rot
Penicillium spp.		Tuber soft rot
Pestalotiopsis cruenta (Syd.) Stey.		Leaf spot
Phaeoramularia dioscoreae (Ellis & Mart.) Deight.	*Cercospora dioscoreae* Ellis & Martin	Leaf spot
Phakopsora dioscoreae Thaung		Rust
Phylleutypa dioscoreae (Wakef.) Petr.		Witches' broom
Phyllosticta dioscoreae (Cooke) Cooke		Leaf spot
Phyllosticta dioscoricola P. Brun		Leaf spot
Phyllosticta grafficina Sacc.		Leaf spot
Pseudocercospora contraria (H. & P. Syd.) Deighton	*Cercospora contraria* Syd.	Leaf spot
Rhizopus nodosus Namyslowski		Tuber soft rot
Rhizopus sp.		Tuber soft rot
Rosellinia bunodes (Berk. & Br.) Sacc.		Tuber soft rot
Serratia sp.		Tuber soft rot
Sphaerostilbe repens Berk. & Br.		Tuber rot
Thanatephorus cucumeris (Frank) Donk.	*Corticium solani* (Prill. & Delacr.) Bourd. & Galz. *Rhizoctonia solani* Kühn	Leaf blight, tuber soft rot
Trichoderma viride Pers. ex S.F. Gray		Tuber soft rot
Uredo dioscoreae P. Henn.		Rust
Uromyces combreti Thaung		Rust

TARO

Species	Synonymy	Common name
Alternaria tenuissima (Kunze ex Pers.) Wilts.		Leaf spot
Botryodiplodia theobromae Pat.		Corm rot

Species	Synonymy	Common name
Botrytis spp.		Corm rot
Cercosporidium sp.		
Cladosporium colocasiae Saw.		Brown leaf spot
Corticium rolfsii Curzi	*Sclerotium rolfsii* Sacc.	*Sclerotium* rot
Diplodia sp.		Black rot or Java black rot
Endoconidiophora sp.		Black rot
Erwinia chrysanthemi Burkholder, McFadden and Dimock		Soft rot
Erwinia sp.		Soft rot, corm rot
Fusarium oxysporum Schlecht.		Soft rot
Fusarium solani (Mart.) Sacc.		Dry rot
Fusarium spp.		Corm rot
Johnstonia colocasiae M.B. Ellis		Leaf spot
Leptosphaerulina trifolii (Rostr.) Petrak		Leaf spot
Macrophomina phaseolina (Tassi) Goid	*Rhizoctonia bataticola* (Taub.) Butler	Leaf rot
Myrothecium roridum Tode ex Fr.		
Phyllosticta colocasiicola Höhn		Shothole leaf spot
Phytophthora cinnamomi Rands		Corm rot
Phytophthora colocasiae Racib.		Leaf blight, corm rot of eddoe
Phytophthora nicotianae var. *parasitica* (Dastur) Waterhouse		Corm rot
Phytophthora palmivora (Butler) Butler		Leaf blight
Pleosphaerulina sp.		
Pseudocercospora colocasiae Deighton		Leaf spot
Pythium myriotylum Dreschl.		Root rot
Pythium splendens Braun		Root rot
Pythium spp.		Corm rot
Rhizoctonia sp.		Corm rot
Thanatephorus cucumeris (Frank) Donk	*Rhizoctonia solani* Kühn *Corticium solani* (Prill. & Delacr.) Bourd. & Galz.	

CHECK LIST OF NEMATODES

Species	Common name
Aphelenchoides besseyi Christie, 1942	
Belonolaimus sp.	
Criconemoides sp.	
Ditylenchus destructor Thorne, 1945	
Ditylenchus dipsaci (Kühne, 1857) Filipjev, 1936	
Helicotylenchus erythrinae (Zimm., 1904) Golden, 1956	
Helicotylenchus sp.	
Hemicycliophora spp.	
Hoplolaimus pararobustus (Schuur. Stek. & Teun., 1938) Sher, 1963	
Hoplolaimus sp.	
Longidorus siddiqii Aboul-Eid, 1971	
Longidorus sp.	
Macroposthonia sp.	
Meloidogyne arenaria (Neal, 1889) Chitwood, 1949	Root-knot nematodes
Meloidogyne hapla Chitwood, 1949	
Meloidogyne incognita (Kofoid & White, 1919) Chitwood, 1949	
Meloidogyne javanica (Treub, 1885) Chitwood, 1949	
Paratylenchus sp.	
Peltamigratus nigeriensis Sher, 1964	
Pratylenchus brachyurus (Godfrey, 1929) Filipjev & Shuur. Stek., 1974	
Pratylenchus coffeae (Zimm, 1898) Filipjev & Shuur. Stek., 1941	Lesion nematode
Pratylenchus sp.	
Radopholus similis (Cobb, 1893) Thorne, 1949	
Rotylenchulus reniformis Linford & Oliviera, 1940	Reniform nematode
Rotylenchulus sp.	
Scutellonema bradys (Steiner & Letlew, 1933) Andrássy, 1958)	Yam nematode
Scutellonema sp.	
Tetylenchus sp.	
Trichodorus porosus Allen, 1957	
Trichodorus sp.	
Tylenchorhynchus sp.	
Tylenchus sp.	

Species	Common name
Xiphinema insignis Loos, 1949	
Xiphinema spp.	

CHECK LIST OF INSECTS AND MITES

SWEET POTATO

Insect	Some synonyms or misidentifications	Common name
Agrius convolvuli (L.)	*Herse convolvuli* (L.)	Sweet potato hornworm, convolvulus hawkmoth
Alcidodes dentipes (Ol.)		Striped sweet potato weevil
Alcidodes erroneus (Thoms)	*Alcidodes orientalis* (Chevr.)	
Alcidodes fabricii (F.)		
Aphis citricola Goot		
Aphis gossypii Glov.		Cotton aphid, melon aphid
Aspidomorpha adhaerens (Weber)		Tortoise beetle
Aspidomorpha australasiae Boisd.		Tortoise beetle
Aspidomorpha dissentanea Boh.		Tortoise beetle
Aspidomorpha miliaris (F.)		Tortoise beetle
Aspidomorpha quadriradiata Boh.		Tortoise beetle
Aspidomorpha socia Montr.		Tortoise beetle
Bedellia ipomoeae Bradley		Sweet potato leaf miner
Bedellia somnulentella (Zell.)		Sweet potato leaf miner
Bemisia tabaci (Genn.)		Whitefly
Brachmia convolvuli (Wlsm.)		Leaf folder
Brachmia macroscopa Meyr.		Leaf folder
Cicadella viridis (L.)		
Conchyloctaenia punctata (F.)		Tortoise beetle
Cylas brunneus F.		Sweet potato weevil
Cylas formicarius (F.)		Sweet potato weevil
Cylas puncticollis Boh.		Sweet potato weevil
Diacrisia obliqua (Wlk.)		Bihar hairy caterpillar
Euchromia polymena (L.)		
Euscepes postfasciatus (Fairm.)		Scarabee weevil, West Indian sweet potato weevil

Insect	Some synonyms or misidentifications	Common name
Hippotion celerio (L.)		Taro hawkmoth, silver striped hawkmoth
Megastes grandalis Gn.		Sweet potato pyralid moth, stem borer
Metriona circumdata (Hbst.)	*Cassida circumdata* Hbst.	Tortoise beetle
Macrosiphum solanifolii (Ashm.)		
Myzus persicae (Sulz.)		
Omphisa anastomosalis (Gn.)		Sweet potato stem or vine borer
Orosius ryukyuensis (Ishihara)	*Nesophrosyne ryukyuensis* (Ishihara)	
Pilocrocis sp.		
Swinhoea vegetus (Swinh.)	*Cretonia vegeta* Swinh.	
Synanthedon dasysceles Bradley		Clearwing moth
Synanthedon leptosceles Bradley		Clearwing moth
Tabidia aculealis (wlk.)		Leaf folder
Trialeurodes abutilonea (Haldeman)		Whitefly

CASSAVA

Insect	Some synonyms or misidentifications	Common name
Acromyrmex sp.		Leaf-cutting ant
Ahasverus advena (Waltl.)		Foreign grain beetle
Aleurotrachelus sp.		Whitefly
Aleurothrixus aepim (Göldi)	*Aleurothrixus graneli* Blanchard	Whitefly
Alphitobius spp.		Lesser mealworm beetles, black fungus beetles
Anastrepha manihoti Costa Lima		Cassava fruitfly
Anastrepha pickeli Costa Lima		Cassava fruitfly
Aonidomytilus albus (Ckll.)	*Lepidosaphes alba* (Ckll.)	Cassava stem mussel scale
Araecerus fasciculatus (Deg.)		Coffee bean weevil, cacao weevil
Atta sp.		Leaf-cutting ant
Bemisia hancocki Corbett		Whitefly
Bemisia tabaci (Genn.)		Whitefly

Insect	Some synonyms or misidentifications	Common name
Bostrychoplites cornutus (Ol.)		
Carpolonchaea chalybea (Wied.)	*Lonchaea chalybea* Wied.	Shoot fly
Carpophilus spp.		Dried fruit beetles
Cathartus quadricollis (Guér.)		Square necked grain beetle
Ceroplastes cirripediformis Comst.		
Chilomina clarkei (Amsell)		
Coelosternus manihoti Mshl.		
Coninomus constrictus (Gyl.)		
Corynothrips stenopterus Williams		
Crytolestes spp.		Rust red and flat grain beetles
Dermestes spp.		Hide beetles
Dinoderus spp.		
Ephestia cautella (Wlk.)		Tropical warehouse moth
Erinnyis ello (L.)	*Dilophonata ello* L.	Cassava hornworm
Ferrisia virgata (Ckll.)		
Gnathocerus cornutus (F.)		Broadhorned flour beetle
Heterobostrychus brunneus (Murr.)		
Lasioderma serricorne (F.)		Cigarette beetle
Leucopholis coneophora Burm		White grub
Necrobia rufipes (Deg.)		Red legged ham beetle, copra beetle
Oryzaephilus mercator (Fauv.)		Merchant grain beetle
Oryzaephilus surinamensis (L.)		Saw toothed grain beetle
Palorus ratzeburgi (Wissm.)		Small eyed flour beetle
Parasaissetia nigra (Nietn.)	*Saissetia nigra* (Nietn.)	
Phenacoccus gossypii Tns. & Ckll.		
Phenacoccus manihoti Matile-Ferrero		
Plodia interpunctella (Hbn.)		Indian meal moth
Polistes canadensis L.		
Pseudaulacaspis pentagona (Targ.)		White peach scale
Pyralis manihotalis (Gn.)		
Retithrips syriacus (Mayet)		

Insect	Some synonyms or misidentifications	Common name
Rhyzopertha dominica (F.)		Lesser grain borer
Saissetia coffeae (Wlk.)	*Saissetia hemispherica* (Targ.)	
Saissetia miranda (Ckll. & Parr.)		
Scirtothrips manihoti (Bondar)	*Euthrips manihoti* Bondar	
Silba pendula Bezzi		
Sitophilus oryzae (L.)		Lesser rice weevil
Stegobium paniceum (L.)		Biscuit beetle, drug store beetle
Tenebroides mauritanicus (L.)		Cadelle
Thaneroclerus buquetti (Lef.)		
Trialeurodes manihoti (Bondar)		
Trialeurodes rara Singh		
Tribolium castaneum (Hbst.)		Rust red flour beetle
Trichogramma fasciatum (Perkins)		
Vatiga manihotae (Drake)	*Leptopharsa manihotae* Drake	Cassava lacebug
Zonocerus variegatus (L.)		Variegated grasshopper
Mite		
Eutetranychus orientalis (Klein)		
Mononychellus mcgregori Flechtmann & Baker		
Mononychellus tanajoa (Bondar)	*Tetranychus tanajoa* (Bondar) *Mononychus tanajoa* (Bondar)	Tanajoa, green cassava mite, green spider mite
Oligonychus (Homonychus) peruvianus (McGregor)		
Oligonychus (Pritchardinychus) biharensis (Hirst)		
Oligonychus (Reckiella) gossypii (Zacher)		
Tetranychus (Tetranychus) cinnabarinus (Boisd.)	*Tetranychus (Tetranychus) telarius* L. *Tetranychus (Tetranychus) urticae* Koch	Spider mite, carmine spider mite
Tetranychus (Tetranychus) marianae McGregor	*Tetranychus evansi* Baker & Pritchard	

YAMS

Insect	Some synonyms or misidentifications	Common name
Aphis gossypii Glov.		Cotton aphid
Aspidiella hartii (Ckll.)		Yam scale
Geococcus coffeae Green		Coffee root mealybug
Gymnogryllus lucens (Wlk.)	*Gryllus lucens* Wlk.	Cricket
Heteroligus appius (Burm.)		Lesser yam beetle
Heteroligus meles (Billb.)	*Heteroligus claudius* Klug	Greater yam tuber beetle, yam beetle
Lilioceris livida (Dalm.)	*Crioceris livida* Dalm. *Crioceris crampeli* Pic.	
Palaeopus costicollis Mshl.	*Palaeopus dioscoreae* Pierce	Yam weevil
Phenacoccus gossypii Tns. & Ckll		
Planococcus citri (Risso)	*Pseudococcus citri* (Risso)	Citrus mealybug
Planococcus dioscoreae Williams		
Prionoryctes caniculus Arr.		Yam beetle
Prionoryctes rufopiceus Arr.		Yam beetle
Senoclidia purpurata (F.Sm.)	*Senoclidia furva* Konow *Senoclidia violaceipennis* Cameron	Sawfly

TARO

Insect	Some synonyms or misidentifications	Common name
Agrius convolvuli (L.)	*Herse convolvuli* (L.)	Sweet potato hornworm, convolvulus hawkmoth
Aphis gossypii Glov.		Cotton aphid, melon aphid
Bemisia tabaci (Genn.)		Whitefly
Campsomeris dorsata (F.)		
Cyrtorhinus fulvus Knight		
Hippotion celerio (L.)		Taro sphinx moth, taro hornworm, silver striped hawkmoth
Ligyrus cuniculus (F.)		
Ligyrus ebenus (Degeer)		
Myzus persicae (Sulz.)		
Papuana huebneri (Fairm)	*Pimelopus hübneri* Fairm. *Papuana hübneri* Arrow	Taro beetle

Insect	Some synonyms or misidentifications	Common name
Papuana inermis Prell		Taro beetle
Papuana woodlarkiana (Montr.)	*Scarabaeus woodlarkianus* Montr.	Taro beetle
Pentalonia nigronervosa Coq.		
Planococcus citri (Risso)	*Pseudococcus citri* (Risso)	
Pseudococcus longispinus (Targ.)	*Pseudococcus adonidum* (L.)	
Spodoptera litura (F.)	*Prodenia litura* (F.)	Cotton leafworm, taro cutworm
Tarophagus proserpina (Kirk)	*Megamelus proserpina* Kirk	Taro leafhopper

Mite		
Tetranychus (Tetranychus) marianae McGregor	*Tetranychus evansi* Baker & Pritchard	

PESTICIDE CHECK LIST

The inclusion of a trade name does not indicate approval of that product.
F = fungicide, H = herbicide, I = insecticide, N = nematicide, R = rodenticide

Common name	Trade names
Alachlor H	Lasso
Aldicarb IN	Temik
Aldrin I	Aldrex and numerous others
Aluminium phosphide R	Phostoxin, Celphos, Delicia
Ametryne H	Primatol
Atrazine H	Numerous trade names
Bacillus thuringiensis (Berliner) I	Dipel, Thuricide
Benomyl F	Benlate
γ-BHC (lindane) I	Bexa dust, Gammalin and numerous others
BHC + carbaryl I	Sevidol
Binapacryl AF	Morocide
Bordeaux mixture F	copper sulphate and hydrated lime solution
Butylate H	Sutan
Calciferol + warfarin R	Sorexa CR
Captan F	Orthocide
Carbaryl I	Dicarbam, Sevin
Carbofuran IN	Furadan
Carbofenothion I	Trithion
Carboxylic acid F	Luprosyl
Chloramben H	Amiben, Vegiben

Common name	Trade names
Chlorbromuron H	Maloran
Chlorfenvinphos I	Birlane
Chlorobenzilate A	Akar, Benzilan, Folbex
Chlorophacinone R	Drat
Chlorthal-dimethyl H	Dacthal
Copper oxychloride F	Perekol, Hi-Bordeaux and many others
Copper sulphate F	Cutrine
Coumachlor R	Tomorin
Coumatetralyl R	Racumin
Dalapon H	Dowpon, Basfapon, Gramevin
D-D N	D-D
DDT I	Numerous trade names
Demeton-methyl I	Metasystox
Diazinon I	Basudin, Neocide and many others
Dichlorvos I	Nogas, Nuvan
Dicofol I	Kelthane
Dieldrin I	Numerous trade names
Difenacoum R	Neosorexa, Ratak
Dimethoate I	Perfekthion, Rogor
Diphacinone R	Diphacin
Diphenamid H	Dymid, Enide
Disulfoton I	Solvirex
Diuron H	Numerous trade names
EDB (ethylene dibromide) N	Dowfume, Nemafume
Endosulfan I	Thiodan
EPTC H	Eptam
Fenitrothion I	Sumithion, Folithion
Fensulfothion N	Dasanit, Terracur-P
Fenthion I	Lebaycid
Ferbam F	Trifungol, Hokmate and many others
Fluometuron H	Cotoran
Glyphosate H	Roundup
Linuron H	Afalon
Malathion I	Cythion and many others
Mancozeb AF	Dithane M 45
Maneb F	Manzate, Dithane M 22
Mephosfolan I	Cytrolane
Methamidophos I	Tamaron
Methazole H	Paxilon, Probe
Methomyl I	Lannate
Metobromuron H	Patoran
Metribuzin H	Lexone, Sencor
Monocrotophos I	Azodrin, Nuvacron
Monolinuron H	Gramonol, Aresin
MSMA H	Daconate and many others
Naled AI	Dibrom
Napropamide H	Devrinol
Nitrofen H	NIP, TOK
Nitralin H	Planavin
Noruron H	Herban
Omethoate I	Folimat

Common name		Trade names
Oxadiazon	H	Ronstar
Paraquat	H	Gramoxone
Parathion	I	Rhodiatox
Phorate	I	Thimet
Phosalone	I	Zolone
Phosphamidon	I	Dimecron
Prometryne	H	Gesagard, Caparol
Propanil, DCPA (Japan)	H	Stam, Surcopur
Ethoprophos	N	Prophos, Mocap
Propineb	F	Antracol
Quinalphos	I	Ekalux, Bayrusil
Simazine	H	Gesoran, Simadex and many others
Sodium cyanide	R	Cymag
TCA	H	NaTA
Thiabendazole	F	Mertect, TBZ
Thiometon	I	Ekatin
Trichlorphon	I	Dipterex
Trifluralin	H	Treflan
Vernolate	H	Bexide, Vernam
Warfarin	R	Numerous trade names
White Oil	I	Refined grades of petroleum oils, Volck
Zinc phosphate	R	Phosvin
Zineb	F	Numerous trade names
Ziram	F	Numerous trade names

INDEX

C

D

N

O

P

Q

R

S

T